What is Global Engineering Education For?

The Making of International Educators

Parts I and II

Synthesis Lectures on Global Engineering

Editor
Gary Downey, *Virginia Tech*
Assistant Editor
Kacey Beddoes, *Virginia Tech*

The Global Engineering Series challenges engineering students and working engineers to cross the borders of countries, and it follows those who do. Engineers and engineering have grown up within countries. The visions engineers have had of themselves, their knowledge, and their service have varied dramatically over time and across territorial spaces. Engineers now follow diasporas of industrial corporations, NGOs, and other transnational employers across the planet. To what extent do engineers carry their countries with them? What are key sites of encounters among engineers and non-engineers across the borders of countries? What is at stake when engineers encounter others who understand their knowledge, objectives, work, and identities differently? What is engineering now for? What are engineers now for? The Series invites short manuscripts making visible the experiences of engineers and engineering students across the borders of countries. Possible topics include engineers in and out of countries, physical mobility and travel, virtual mobility and travel, geo-spatial distributions of work, international education, international work environments, transnational identities and identity issues, transnational organizations, research collaborations, global normativities, and encounters among engineers and non-engineers across country borders. The Series juxtaposes contributions from distinct disciplinary, analytical, and geographical perspectives to encourage readers to look beyond familiar intellectual and geographical boundaries for insight and guidance. Holding paramount the goal of high-quality scholarship, it offers learning resources to engineering students and working engineers crossing the borders of countries. Its commitment is to help them improve engineering work through critical self-analysis and listening.

What is Global Engineering Education For? The Making of International Educators – Parts I and II
Gary Lee Downey and Kacey Beddoes
2011

What is Global Engineering Education For? The Making of International Educators – Parts I and II

Gary Lee Downey and Kacey Beddoes

ISBN: 978-3-031-00996-9   paperback
ISBN: 978-3-031-02124-4   ebook

DOI: 10.1007/978-3-031-02124-4

A Publication in the Springer series
SYNTHESIS LECTURES ON GLOBAL ENGINEERING

Lecture #1
Series Editor: Gary Downey, *Virginia Tech*
Assistant Editor: Kacey Beddoes, *Virginia Tech*
Series ISSN
Synthesis Lectures on Global Engineering
ISSN pending.

What is Global Engineering Education For?

The Making of International Educators

Parts I and II

Gary Lee Downey and Kacey Beddoes
Virginia Tech

SYNTHESIS LECTURES ON GLOBAL ENGINEERING #1

ABSTRACT

Global engineering offers the seductive image of engineers figuring out how to optimize work through collaboration and mobility. Its biggest challenge to engineers, however, is more fundamental and difficult: to better understand what they know and value *qua* engineers and why. This volume reports an experimental effort to help sixteen engineering educators produce "personal geographies" describing what led them to make risky career commitments to international and global engineering education. The contents of their diverse trajectories stand out in extending far beyond the narrower image of producing globally-competent engineers. Their personal geographies repeatedly highlight experiences of incongruence beyond home countries that provoked them to see themselves and understand their knowledge differently. The experiences were sufficiently profound to motivate them to design educational experiences that could challenge engineering students in similar ways.

For nine engineers, gaining new international knowledge challenged assumptions that engineering work and life are limited to purely technical practices, compelling explicit attention to broader value commitments. For five non-engineers and two hybrids, gaining new international knowledge fueled ambitions to help engineering students better recognize and critically examine the broader value commitments in their work.

A background chapter examines the historical emergence of international engineering education in the United States, and an epilogue explores what it might take to integrate practices of critical self-analysis more systematically in the education and training of engineers. Two appendices and two online supplements describe the unique research process that generated these personal geographies, especially the workshop at the U.S. National Academy of Engineering in which authors were prohibited from participating in discussions of their manuscripts.

KEYWORDS

engineering education, global engineering (education), international engineering (education), study abroad, international service learning, international co-op, international internship, work abroad, Peace Corps, ABET, EC 2000, globalization

For my students, who give my work meaning

Gary Downey

For my mentors, who have given me more than they know

Kacey Beddoes

Contents

PART I

Introduction

CHAPTER 1

The Border Crossers: Personal Geographies of International and Global Engineering Educators

Gary Lee Downey

I'm so frustrated by always having to explain why we should be doing this. Not how and the details, but why is this important?

Les Gerhardt, Rensselaer Polytechnic Institute

You had an enlightening moment in your paper when you said "We should do that." Many people say that. Most people don't. You did. What is it about your background, your makeup, your personal set of convictions that aligned you with the situation and made you simply plunge forward with incredible investment of time and effort?

John Grandin, University of Rhode Island,
commenting on Jim Mihelcic's manuscript

"What led a bunch of crazy people to devote so much of their time and energy to helping students gain new forms of knowledge that might be called 'international'?" This was the challenge Kacey Beddoes and I posed to the authors of sixteen draft "personal geographies" about their work in international engineering education. They were preparing revisions of their manuscripts following a two-day workshop at the U.S. National Academy of Engineering. We could have referred to "global" knowledge rather than international knowledge, but we wanted to avoid simply replicating the pressure they were all feeling to position their work in relation to the newly dominant image of "global engineering education." Rather, we wanted them to rethink and reframe their descriptions of their career trajectories in light of what they heard and said in interaction with one another. The distinction between international and global knowledge in engineering education (or the relationship between the two) was one live issue on the table for everyone.

This volume reports an experimental effort to develop and share a process for challenging a unique and diverse group of U.S. engineering educators to examine and report on their intellectual and career trajectories in relation to one another. These contributors stand out in that all are focused on a particular set of alternative practices I call here "international and global engineering education," using "and" to keep active the ambiguity in the distinction.

The sixteen personal geographies in this volume describe the distinct pathways that led these educators to make risky career commitments to international and global engineering education. The trajectories contributors trace stand out in extending far beyond the narrowly-defined image of people preparing themselves to produce "globally-competent" engineers. Their accounts reveal key moments in which they began to question and reflect on their own knowledge and broader norma-tive commitments. They describe how what I call here "adding identities outside home countries" provoked them to recognize and reflect critically on identities and practices they had previously taken for granted.

These personal geographies also describe how the surprises, confusions, worries, and plea-sures of adding identities beyond the borders of home countries led contributors to want others, in particular prospective engineers, to experience similar challenges. Like many others committed to international and global engineering education, these sixteen educators invested huge amounts of time and energy to redefine what is included in the training of engineers and, ultimately, engineering practices. Their messages and practices are not uniform, nor even consistent with one another. Both the overlaps and the differences among them warrant further discussion and analysis. Still, each holds in his or her own way that learning to reflect critically on their knowledge and normative commitments should be a key practice in the education of engineers. Together they maintain that international and global engineering education can provide crucial, epistemologically-significant sites for transformative learning. And the compelling evidence came from life experiences outside home countries.

MAPPING THE PRESENT IN INTERNATIONAL AND GLOBAL ENGINEERING EDUCATION

During the decade prior to initiating this project in 2007, I had long been impressed by the con-viction with which advocates of international and global engineering education had characterized their work (at national meetings, workshops, etc.). Many exhibited confidence that educational ex-periences in other countries through such activities as study abroad, internships, and service learning provided forms of knowledge that could help engineering students become practitioners with an ap-preciation for difference, tolerance for ambiguity, and acute understanding of their own limitations. Some advocates had clearly made substantial career sacrifices, departing dominant tracks of career advancement and accepting comparatively marginal trajectories developing and managing educa-tional programs. Trained as both mechanical engineer and cultural anthropologist (and a committed educator of engineers myself), I found myself asking what it was about this work in international and global engineering education that made it so compelling to these people?

The same period was also witness to the emergence of another phenomenon: international and global education began to gain substantial visibility both within engineering communities and beyond them. This new visibility was especially interesting in the case of engineering because the force of its emergence appeared to lie almost entirely beyond the agency and control of long-standing practitioners. The justifications I began hearing for international and global engineering education often did not fit with what I had been hearing from them. For me this raised questions about the relationship between the two, as well as about likely future courses of development for international and global engineering education in the United States. In particular, what broader social projects have they been and will they be serving?

As Brent Jesiek and Kacey Beddoes explain in Chapter 2, the image of international and global education that has scaled up to dominance across engineering communities is "education for global competitiveness." The image of competitiveness emerged during the 1980s to refer to the economic relationships among countries. In this image, preparing engineers for international work is a crucial practice in a multi-sectoral (industry-university-government) effort to advance American economic competitiveness in a world in which private industry has gone multinational in scope. The broad image of economic competitiveness has arguably provided the dominant U.S. frame for planetary relations since the decline and end of the Cold War.[1] For engineering educators to expect to link engineering education to economic competitiveness is not surprising since educators have been adjusting curricula to fit the broader, evolving goal of low-cost production for mass consumption since the 1870s.[2]

Since 2000, the image of education for competitiveness has also come to refer (both inside and outside the United States) to the career challenges of individual engineers participating in the globalization of industry. If industry has gone multinational, then individual engineers must gain "global competence" in order to successfully locate employment and build careers.[3]

By 2004, the National Science Foundation (NSF) had officially embraced the image of economic competitiveness as grounding organizational mandates. A letter from the NSF leadership to the President's Council of Advisors on Science and Technology, for example, characterized the goal of economic competitiveness as economic "supremacy" in a newly leveling world. "Civilization is on the brink of a new industrial order," the letter asserted. "The big winners in the increasingly fierce

[1]Growing acceptance of the image of economic competitiveness was linked to increased focus on the stock market price of U.S.-based private companies as the main indicator of competitive performance. Some relevant early reports include: American Society for Engineering Education, "A National Action Agenda for Engineering Education," 1987; MIT Commission on Industrial Productivity, "Education and Training for the United States," 1989; National Science Board, "The Role of the National Science Foundation in Economic Competitiveness," 1988; The National Commission on Excellence in Education, "A Nation at Risk," 1983. For early critical analysis of the linkage between economic competitiveness and social progress, see Goldman, *Science, Technology, and Social Progress*, 1989. For analysis of the image of economic competitiveness among engineers, see Downey, "CAD/CAM Saves the Nation?" 1992; Downey, "The World of Industry-University-Government," 1995; Downey, *The Machine in Me*, 1998a. For analysis of the image of economic competitiveness in NSF policy, see Lucena, *Defending the Nation*, 2005.

[2]Downey, "Low Cost, Mass Use," 2007.

[3]The 2006 report, *In Search of Global Engineering Excellence: Educating the Next Generation of Engineers for the Global Workplace*, sponsored and published by Continental AG in Hanover, Germany, brought together educational leaders from eight prestigious engineering universities in Brazil, China, Germany, Japan, Switzerland, and the United States. The new "globally prepared engineers" asserts the Executive Summary, must have "[t]he ability to live and work in a global community," including "broad engineering skills and know-how" and the ability "to be flexible and mobile, and able to work internationally" (p. 1). For work focused on the question of global competence, see the *The Online Journal for Global Engineering Education* (http://digitalcommons.uri.edu/ojgee/).

global scramble for supremacy," it continued, "will not be those who simply make commodities faster and cheaper than the competition." Rather, "[t]hey will be those who develop talent, techniques, and tools so advanced that there is no competition."[4]

The image linking global engineering education to economic competitiveness gained particular force in official engineering circles following the 2005 publication of Thomas Friedman's *The World is Flat: A Brief History of the Twenty-First Century.*[5] Within a year I had lost count of the number of engineers who had told me (and whom I had watched tell one another) that Friedman's image of a flat world had profoundly changed their view of the world (or better, their view of themselves in it). In 2007 the National Academy of Engineering, National Academy of Sciences, and Institute of Medicine together invoked Friedman to introduce their *Rising Above the Gathering Storm: Energizing and Employing America for a Brighter Economic Future.* "[T]he international economic playing field is now 'more level' than it has ever been," their report echoed, and the United States is decidedly not "investing in our future and preparing our children the way we need to for the race ahead."[6]

The National Academy of Engineering had already in 2005 named the stakes in competitiveness for individual engineering careers. "[E]ngineering will only contribute to success," it asserted, "if it is able to adapt to new trends and provide education to the next generation of students so as to arm them with the tools needed for the world as it will be, not as it is today."[7] Engineers will have to be able to communicate not only with technologies but also with customers, it continued, for mass production is giving way to "a buyer-centric business strategy that combines mass customization with customized marketing."[8] And since engineers will be working increasingly in international, interdisciplinary teams, they must achieve "an understanding of the complexities associated with a global market and social context."[9]

This project began in the midst of these developments. In a long telephone conversation in 2007, Dr. Russell Pimmel, senior program officer at NSF, expressed interest in scaling up participation in global engineering education. Dr. Pimmel described how many small programs devoted to global engineering education had sprung up in engineering schools across the country. A few were making serious efforts to expand their purview. "How can we increase the scale of these programs?" he asked. "The level of participation remains very low."[10]

Both Dr. Pimmel and I knew that what was taking place in the small programs involved more than preparing students to follow employers to other countries, even if the image of competitiveness could often be found in those programs. It was also clear that practitioners of study abroad, student exchanges, work internships, and service learning did not portray their objectives and practices in uniform terms, with the possible exception of the assertion that the experience was life-changing.

[4]Committee on Prospering in the Global Economy of the 21st Century, *Rising above the Gathering Storm*, 2007.
[5]Friedman, *The World Is Flat*, 2005.
[6]Committee on Prospering in the Global Economy of the 21st Century, *Rising above the Gathering Storm*, 2007.
[7]National Academy of Engineering, *Educating the Engineer of 2020*, 2005.
[8]National Academy of Engineering, *Educating the Engineer of 2020*, 2005.
[9]National Academy of Engineering, *Educating the Engineer of 2020*, 2005. A number of critiques of links between engineering and competitiveness have also appeared. See Baillie, *Engineers within a Local and Global Society*, 2006; Catalano, *Engineering Ethics, Peace, Justice, and the Earth*, 2006; Riley, "Resisting Neoliberalism," 2007; Riley, *Engineering and Social Justice*, 2008.
[10]Downey notebook 2007 -1:86.

Indeed, some practitioners exhibited strong feelings of resistance to the image of education for competitiveness. In the early stages of this project, one told me s/he had started off in international education with a passion for helping students examine and challenge their own boundaries but was now feeling s/he was "selling a product."[11] I had also frequently sensed that some had begun using the word "global" as a strategic device for winning institutional legitimacy and support while striving to achieve more diverse local goals. I had indeed sometimes done that myself.

I suggested to Dr. Pimmel that the most immediate question was not to figure out how to scale up international and global engineering education as quickly as possible. The immediate question was to figure out what is at stake in international and global engineering education in the first place, and for whom. Figuring out what is at stake and for whom is a prerequisite to identifying, debating, and advocating specific directions and practices. What emerges over time in international and global engineering education will be a product of the actual range of trajectories interested participants select—from emphasizing a particular educational practice to dropping the whole idea entirely. I had seen disconnects between official advocacy of education for global competitiveness and a much more diverse set of actual practices in international and global engineering education. This suggested there was much to explore.

My proposal was to "map the present" in international and global engineering education in the United States. How and why have practitioners of international and global engineering education become interested in such work? How have educators built relations with others, both locally and elsewhere, who may or may not have been interested in their projects? What practices have they developed, with what contents, and to what ends? What efforts have they made to integrate their initiatives into established practices of engineering formation? Mapping the present could reveal how educators differentially embrace, strategically deploy, or avoid the image of "global engineering education" that had risen to dominance over the image of "international engineering education" that had long been dominant among practitioners.[12]

I could have proposed an ethnographic study critically examining contemporary struggles over international and global engineering education.[13] The resulting volume would have claimed to be an authoritative ethnographic account attempting to inform those struggles by offering findings and arguments and hoping these would diffuse outward by persuading readers of their value.[14]

But that was not the pathway I wanted to try here. The grant category Dr. Pimmel and I were discussing was specifically to support workshops, not individual research. I took this constraint as a challenge and an opportunity to produce a different kind of research process with a different kind of writing and analysis as its outcome. The challenge would be to figure out how to enroll practitioners of international and global engineering education as researchers of their own trajectories. The project

---

[11] Confidential conversation, October 2007.

[12] Examining the shift from one to the other is the justification for Chapter 2 by Brent Jesiek and Kacey Beddoes.

[13] As Jesiek and Beddoes explain, their historical overview is meant to be a starting point. Relying on the published arguments of advocates rather than in-depth interviews, it identifies dominant trends and categories to help readers appreciate that the initiatives described here all had antecedents. Each new initiative gained significance in relation to established categories with which it had to compete for meaning.

[14] For an early critique of the "diffusion model" of knowledge dissemination, see Latour, *Science in Action*, 1987.

would be, in the first instance, to help these educators make visible, both to themselves and to others, the diversity of pathways they have followed. It would be to persuade them to analyze and share with others the development of their knowledge and commitments, to help reveal what might otherwise remain hidden. The analysis of educational practitioners would be done by the practitioners themselves.[15]

PERSONAL GEOGRAPHIES TO MAP DIFFERENCES

The contributors to this volume offer you "personal geographies" of their pathways through international and global engineering education.[16] A personal geography is a narrative map of a trajectory. When geographers write personal geographies, they typically mean personal experiences of trajectories through physical spaces, especially to highlight experiences hidden by dominant images of those spaces, such as the home.[17] This project asked participants to use the idea of trajectory metaphorically. While it certainly refers to movements through physical space (e.g., travels to India, Bolivia, California, Atlanta), it also refers to movements through institutional locations (e.g., faculty member in engineering or foreign languages, education administrator, department head, dean) and other identity relationships (e.g., collegial relations, relations with host individuals and organizations). The term "personal geography" challenged writers in this volume to map their changing identities as they have traveled through work and life.

As contributors describe these travels, they report key transitions in which they found themselves in new relationships with unfamiliar people and things, organizations and institutions. I describe these transitions as "adding identities." For the purposes of this project, the term "identities" refers simply to relationships among entities (id-entity). To help identify what made a particular step significant, contributors present themselves as knowledge workers engaged in changing relationships with other knowledge workers. They accepted a challenge to pursue three sets of research questions about themselves and others at key moments: (1) How was I (were they) located? (2) What did I (they) know? (3) What did I (they) want? Think *location*, *knowledge*, and *desire* as a mnemonic device.[18]

Mapping *location* involves examining how one is positioned in relation to others. Not limited to physical geography, location refers broadly to positions of all sorts, e.g., within a career, across an organization, in relation to colleagues. Importantly, it includes the power dimensions of those relationships. It matters greatly, for example, if the educator in question is an adjunct instructor

---

[15]This project benefits from related work on participatory approaches to science and technology, most of which involve expanding the contents of knowledge-based decision making. See for example Lengwiler, "Participatory Approaches in Science and Technology, 2008; Nowotny et al., *Re-Thinking Science*, 2001; Zuiderant-Jerak and Jensen, "Unpacking 'Intervention' in Science and Technology Studies," 2007.

[16]See Appendix 1 for an account of the research project. See Supplements 1 and 2 (http://www.morganclaypool.com/page/downey) for all Workshop documents and evaluation data from the Workshop. Brent Jesiek, Juan Lucena, and Kacey Beddoes served as co-organizers of the Workshop. Thanks to Juan for suggesting the term "personal geography" as a label for these accounts.

[17]Bell, "Insignificant Others," 1991; Cresswell, *In Place/out of Place*, 1996; Silverstone et al., "Information and Communication Technologies and the Moral Economy of the Household," 1992; Sommerville, "Homelessness and the Meaning of Home, 1992. But see the expansion in Valentine, "'Sticks and Stones May Break My Bones'," 1998 .

[18]Downey, "Location, Knowledge, and Desire," 2010a.

or a dean, serving on an engineering or a humanities faculty, employed at a research university or a teaching college. Each of these locations puts one in relationships with some people and things and not others. It also matters if the key moment of transition takes place in relation to colleagues, students, superiors, funders, etc., and, as we will see below, inside or outside the home country.

Mapping *knowledge* involves questioning the forms of knowledge possessed by those one encounters and engages along the way. Making visible forms of knowledge is important because what educators understand as their knowledge helps define what they see or do not see in their work. Advocates of international and global engineering education regularly champion the new knowledge they provide students. It is new knowledge that has warranted calling the experiences life-changing. What does this knowledge include and what does it leave out? A special task in mapping one's trajectory in relation to those of others is to make visible the broad range of forms of knowledge present at key moments. These might include everything from knowledge of contrasting pedagogical and work practices and different identities for engineers and non-engineers across other countries to an understanding of the mechanics of credit transfer processes.

Mapping *desire* involves asking what participants in one's trajectory want. Inquiring into wants helps make visible the "normativities" in different trajectories through international and global engineering education, i.e., the broader, external projects those trajectories aim to serve.[19] It highlights what is at stake in encounters among those with different wants. It is especially useful to examine how desire is related to location and knowledge. What an engineering dean wants from instruction in a foreign language, for example, is unlikely to overlap entirely with what a professor in foreign languages and literatures wants from that instruction. What a student wants in service learning likely does not overlap entirely with what is wanted by those whom students call "locals." Moreover, what any of the above wanted yesterday may be different from what they want today. Desire changes with changing location and knowledge.

The sociologist of knowledge John Law has persuasively argued that research practices in the social sciences miss much in the analysis of social interactions. What they tend to overlook includes, for example, "[p]ains and pleasures, hopes and horrors, intuitions and apprehensions, losses and redemptions, mundanities and visions, angels and demons, things that slip and slide, or appear and disappear, change shape or don't have much form at all, unpredictabilities . . . [and other] [p]arts of the world [that] are . . . vague, diffuse, and indistinct."[20] At the same time, in generating the messiness that is everyday life through contingent actions of various sorts, one also always takes some things for granted, producing patterned persistence in the process. One typically acts in ways, for example, that connect back to something one had done previously. The production of patterned persistence is every bit as much a part of agency as the production of novelty or change. John Law describes studying the messiness of social life, after all, in a book with an introduction and conclusion, an ordered set of chapters, acknowledgements, glossary, notes, references, and index.

---

[19]Methodologically, asking what people want is typically better for revealing normative commitments than asking why they do what they do. The why question imposes a cause-effect framework that focuses attention on finding a cause for an action rather than identifying its directionality.

[20]Law, *After Method: Mess in Social Science Research*, 2004, 2. Thanks to Kacey Beddoes for calling this passage to my attention.

Personal geographies contrast with survey results, summaries of structured interviews, and content analysis of texts (to select a few common research practices) by disclosing or at least gesturing toward some of the messier phenomena of everyday life. Contributors often speak of surprises and other contingencies. They too, however, leave much out. These personal geographies are directed accounts, with directorial authority resting ultimately in the hands of the author. Readers can readily infer that much is left out of every decision, action, or event that an author reports. As historian of technology Rosalind Williams put it in a conversation on the character of personal geographies, "One leaves out a lot in such exercises, so there is more coherence in the telling than in the living." But, she added, "still you can't create a thread if there isn't one."[21]

For me, the key issue in organizing research practices for this project lay not in the extent they would make visible messy contingencies and complexities or call attention to patterned persistence, although both are important and the differences among them are significant. The key issue also lay not in a goal of providing or approaching some sort of complete account of what is now taking place in international and global engineering education. Rather, it lay in what specific strategies promised to make visible or risked hiding in the identities of international and global engineering educators, especially the knowledge and normative contents of those identities.

To prospective participants, the narrative form of the personal geography fell ambiguously between "program history" and "autobiography." The program history was most familiar to them. Its temporal form details the visions, practices, and outcomes of a specific educational initiative, a "program," apart from the agents involved in it. Common at meetings of engineering educators, the program history transforms the activities it reports into a kind of organism that experiences birth, development, and travel through time. The author may appear in the recounting of struggles, but the main protagonist is the program itself, and the narrative becomes its biography. The genre is relatively comfortable for program developers to produce because it ultimately doubles as description and promotion. I have not myself witnessed an educator offering a history of a stillborn, diseased, or otherwise pathological program.

At the other pole, the temporal form of the autobiography narrates many dimensions of an author's life simultaneously. With the author as protagonist, it makes any thought or action fair game for inclusion. Its organizing image is the "life," of which work is but a part. Authors of autobiographies typically present the life as a narrowly-construed mix of willful agency and unanticipated contingency, including what the author says or does or thinks on the one hand and what just happens to the author on the other. Although regularly gesturing toward the messiness of everyday living, it too ultimately finds persistence in some sort of organizing thread, often named in the title or subtitle.

Over the past two decades, anthropologists and other qualitative researchers in the social sciences have experimented broadly with analyses in narrative form that locate the researcher's point of view within the analysis.[22] Including the author in the text always risks reducing analysis to

---

[21] Email message, July 29, 2009.

[22] The literature is vast. Some starting points are Abell, "Narrative Explanation, 2004; Atkinson, *The Ethnographic Imagination*, 1990; Atkinson et al., *Handbook of Ethnography*, 2002; Bruner, "Ethnography as Narrative," 1997; Franzosi, "Narrative Analysis," 1998; Patterson and Monroe, "Narrative in Political Science," 1998.

autobiography, introducing "mere narcissism," as social anthropologist Judith Okely put it in an early reflection on the issue. The reason for this is of course the danger of shifting the focus in the account from the object of study to the author. Yet perhaps most scholars today would agree with Okely's contention that "[s]elf-adoration is quite different from self-awareness and critical scrutiny of the self" and that including the author as a "positioned subject" in the text can add much to it.[23]

Exactly what including the author in the analysis adds or subtracts depends upon the purpose of the project and the specific research and writing strategies one selects. The engineering education researcher Robin Adams and five colleagues have studied "storytelling in engineering education," for example, with the explicit goal of better understanding the emergence of an "engineering education research community." Their focus is, in other words, accounting for an observed convergence and possibly contributing further to it.

They invited eight scholars, including three co-authors, to prepare "story poster" presentations at the national Frontiers in Education conference (supported by two divisions of the IEEE [Institute of Electrical and Electronics Engineers] and one of the American Society for Engineering Education). The organizers asked presenters to respond to a structured set of questions designed to evoke "insider knowledge" pertaining to "driving passions and goals, processes such as getting started and moving forward, difficulties experienced and ways to overcome them, and what they were learning about research."[24] The session used these posters as an anchor for subsequent interactions among the fifty to sixty participants, who posted sticky-note replies onto the posters and engaged in small-group discussions. Session evaluations suggested that the experience helped build a stronger sense of community among participants, and the authors concluded that storytelling "makes explicit our implicit knowledge, promotes reflective practice, and provides entry points into a community of practice."[25]

This study contrasts with the storytelling project in that its goal is explicitly to map differences among contributors by asking all to trace the trajectories of distinct "careers." Like the authors of the story posters, participants in this project responded to questions posed by the organizers. The questions differed not only in inviting reflection over long periods of time (authors did have 10,000 words (originally) to work with, after all), but also in introducing the image of a trajectory as a "sequence of encounters among different perspectives."[26]

As in an autobiography, the temporality of a career matters greatly. What came before what and when did the author acquire which identities? A life is of interest ultimately because of how its various parts stand in relation to one another, perhaps constituting a definitive whole (I recently attended a memorial service for an academic who had been in her nineties in which moving eulogies were delivered not only by colleagues and family members but also by a personal trainer). A career is of interest significantly because it is also a story of others. At every point, it has position in relation to others whom one can see by means of those locations. The words "trajectory" and "geography"

---

[23] Okely, "Anthropology and Autobiography," 1992.
[24] Adams et al., "Storytelling in Engineering Education," 2007 (no pagination).
[25] Adams et al., "Storytelling in Engineering Education," 2007 (no pagination).
[26] See Supplement 1 for workshop guidelines (http://www.morganclaypool.com/page/downey).

emphasize the changing temporal and spatial dimensions of relational location. A personal geography is as much about all those whom the subject has engaged as about the subject her/himself. All become protagonists, even if introduced only briefly.

One key potential limitation in any project mapping differences is incompleteness. Maps make some things visible while hiding others. The knowledge value of a map lies not in its validity or reliability but its plausibility.[27] One cannot judge the accuracy of a road map, for example, directly against reality (validity). One must rely upon other already trusted and accepted exercises in mapping, such as walking the route or invoking GPS satellite technology. It also matters little if the map can be replicated (reliability). The robustness of the collection of maps in this volume depends upon their demonstrable fit with other trusted maps as they chart new terrain. This is what I mean by plausibility.

The volume as a whole can be described as plausibly mapping the territory to the extent it identifies and presents key perspectives involved in international and global engineering education without omitting perspectives whose inclusion would force a redrawing of the terrain. By "perspective" I refer to the landscape one sees by occupying a particular identity. Toward the end of the workshop, one participant announced, "Everyone's here!" to general assent. They meant that a range of identities occupied by international engineering educators was present. Yet it is important to acknowledge that only some positions with stakes in international and global engineering education show up in this volume.

With help from co-organizers, I took care to include the positions of engineering faculty and non-engineering faculty;[28] tenured, untenured, and untenurable; U.S.-born and non-U.S.-born; male and female; white and of color; late, mid, and early-career; senior administration, mid-level administration, and staff; funded and funder. I also wanted to have present diverse commitments to a wide range of external projects. Not included are students, whether participating or non-participating; skeptics of international and global engineering education; donors, both individual and corporate; mediating individuals and organizations between the United States and other countries; or any of the multitude of positions and perspectives outside the United States that participate in U.S. practices of international and global engineering education, from employers of interns to university hosts of study-abroad students to what a number of contributors call "local partners." All these positions must be inferred from the personal geographies. The focus here is educators.

Another key potential limitation in a project mapping differences, including this one, is that it does no more than assert the presence of complexity. The initial force of my reply to Dr. Pimmel's desire to scale up international and global engineering education was, in effect, "It's more complex than that." Yet the claim that one is demonstrating complexity by mapping differences can limit itself by feigning innocence.[29] It can present itself as detached from the very interpretations of the issues it analyzes and hence lacking any responsibilities in relation to them. Kacey Beddoes and I

---

[27] Downey, "Introduction," 1998a, 28.

[28] But see the argument in the Epilogue that all who teach engineers are, in a sense, engineering faculty.

[29] Thanks to Matt Wisnioski for calling my attention to the limitations of feigning innocence, albeit in the work of others rather than our own. See Wisnioski, "Liberal Education Has Failed," 2009.

could, for example, say now to Dr. Pimmel, "Here's a map of what's happening with international and global engineering educators. Now you figure out what to do with it."

But analysis is never innocent because it always bears some relation to dominant practices in the arena under investigation.[30] Certainly one possible relation is no relation: an outcome can be irrelevant. This project explicitly seeks to enhance the relevance of its outcomes for readers interested in international and global engineering education as well as in engineering formation more generally through its techniques for construing careers as trajectories and collaboratively focusing attention on discrepant moments of transition.

The personal geographies in this volume do not straightforwardly present findings. They seek to engage you the reader not only with analyses to assess but also with a collective invitation and challenge.

The invitation is to use these maps to begin examining your own view about international and global engineering education in a new way, as an outcome of adding identities through key transitions of work and life. Contributors found the personal geography to be a new, wholly unfamiliar approach to analysis and writing. Might reading these contributions prompt you to examine when and how you added specific identities that have proven formative in how you regard practices of international and global engineering education? Has adding identities beyond the borders of your home country, in particular, challenged relevant existing professional or personal identities? An hypothesis in this volume is that you, like the contributors, would find it informative to begin assessing how *you* are located, *what* you know, and what *you* want with respect to international and global engineering education. The book is designed not only to share findings but also to elicit responses.

The challenge is to analyze your own knowledge and commitments in (or against) international and global engineering education in relation to those of others and to put yourself in conversation with them, as these contributors have done. Their personal geographies did not emerge in isolation from one another (as Appendix 1 explains) but developed in conversation. Systematically calling attention to differences at the workshop challenged participants to become more knowledgeable about perspectives they had not previously understood or examined. The requirement to actively suggest edits for one another's work further challenged them to try on different perspectives and practice assessing their own trajectories from others' points of view. The focus on differences had the effect of privileging comments that made visible still other perspectives not represented in the room (especially those of collaborating partners inside and outside the United States). And maintaining collaborative conversations through the workshop experience made the work of revision a different kind of writing than rebutting criticism. Participants rewrote their personal geographies with an awareness they were positioning their trajectories in relation to others that differed but nonetheless had value. Virtually every author reported on email how difficult it had been to write, yet many also described the process as exhilarating and uniquely affirming.

The resulting conversation in this volume also conveys a significant finding relative to dominant practices of engineering education: much of the work that has gone into international and

[30]Downey, "What Is Engineering Studies For?" 2009 .

global engineering education is not about adding global competency to engineers as an instrumental skill. Certainly every contributor is struggling to assess how the development of his or her knowledge and commitments relate to the emergent image of education for global competitiveness. Some who accept global competency as an instrumental skill also call attention to additional goals. Some place primary emphasis on alternative goals. Many resist the dominance of competitiveness as a primary, unquestioned justification for international and global engineering education.

The significant news in this book is that these trajectories are challenging dominant practices of engineering education by calling greater attention to the range of stakes in engineering work. Let us turn now to examine how.

REDEFINING ENGINEERING: NINE ENGINEERS

For the nine engineers contributing to this volume, adding identities beyond the borders of home countries challenged assumptions that engineering work, and, hence, life as an engineer are limited to purely technical practices.

1. CALLING ATTENTION TO POLITICAL, SOCIAL, AND HUMAN DIMENSIONS—RICHARD VAZ

Rick Vaz had become a graduate student in electrical engineering at his alma mater, Worcester Polytechnic Institute (WPI), because he had found work as a systems design and product development engineer to be "fundamentally uninteresting."[31] The key addition this move brought to his identity was that of educator. WPI is widely known for its projects-based curricula, which means many people think of project-based learning when they think of WPI but few know how it actually works. "[E]specially interesting and rewarding" to Rick was guiding students through both the senior project in the student's major and the junior project "involve[ing] problems at the interface between society and technology or science." Rick earned his Ph.D. and found a way to stay on at this teaching-oriented institution, winning a tenure-track appointment and ultimately tenure. His identity as engineering professor was secure when a trip to Venice to advise a project measuring tide heights and flow rates initiated a sharp turn.

Rick's account details steps he followed that led ultimately to work pushing students out of what he calls their "comfort zones." This term resonated among workshop participants. It refers to intellectual and personal spaces, and carries an implicit metric of distance. "[M]y colleagues and I are convinced," asserts Rick, "that student learning and growth opportunities are greater when students are brought further out of their comfort zones." The term is significant despite not registering (at all) on the Richter scale of social science theorizing. Everyday learning in engineering certainly carries its discomforts. Rick himself reports he chose electrical engineering because "electricity was a mystery to me." But it is important he experienced learning and work within electrical engineering as something other than pushing himself out of his comfort zone. That was a different discomfort.

---

[31]I quote extensively from the personal geographies below but do not footnote page references. I want to encourage readers to examine a given personal geography closely rather than browse it for highlights.

The widespread use of the term at the Workshop suggested Rick was not alone among engineers in experiencing a unique discomfort himself in working outside the country.

Rick's personal geography reveals a practice of pushing himself out of his comfort zones, adding identities in the process. Desires to become "a diplomat, a therapist, a sociologist, an entrepreneur, or a mentor" were linked to desires to extend his travels as adviser through Washington, D.C. and the Netherlands to Thailand, Namibia, and South Africa. The projects closest to his comfort zone challenged the boundaries of engineering work, for this pushed participants (including especially Rick himself) to face and address "political, social, and human dimensions." With increased distance came enhanced challenges to the person, especially to routine daily practices and expectations. In Thailand, for example, "a central part of our work [was] learning and thinking about culture." These experiences raised questions about his work identity as an engineer.

The increased salience of Rick's identity as an engineer was linked to increased awareness of and commitment to understanding perspectives other than his own. Having plans disintegrate, learning to build trusting relationships, and above all "framing problems based on community perspectives" called attention to the relationship between the work of engineers and the work and lives of others. Rick writes about the "contradictions of international development" and about his students and him rethinking assumptions about "technological advances, social justice, and a sustainable future." He writes about shifting from "addressing problems that have been locally defined and developing solutions that will be locally supported" to establishing more permanent relations with "co-researchers."

Rick's attraction to ever-changing arrays of new questions led him to seek the position of dean of interdisciplinary studies. Rick still carried the identity of tenured engineering professor, but his understanding of engineering work had changed dramatically. Participation in international and global engineering education led him to become and value being many other things in addition to what he had earlier thought was definitive of engineering.

2. LINKING RESEARCH AND EDUCATION TO AN EXPANSIVE LIFE—DAN HIRLEMAN

Dan Hirleman had already come to see himself as a researcher by his junior year of undergraduate studies at Purdue. A key moment was when a German-born professor invited him into his lab. Raised in the Midwest, Dan saw Professor Wittig as "an international." Entering Wittig's lab was entry into an international research community and, with it, acquisition of an international identity. "I was really taken by the fact," Dan writes, "that my engineering work could be, and in fact was, original enough to be of interest to people from around the world."

International research was also a ticket to a "more expansive" life. By this Dan means not only that he wanted his personal identity to complement his work identity as an engineering researcher (and vice-versa). He also wanted each to challenge the other in ways that would keep it stimulated and vibrant, which he feared might not happen.

Dan spent a year in Denmark as a graduate student researcher. Although an unproblematic geographical venture for the international researcher working in an international lab, it was not so for the person. "I was the minority culture," he writes, "and was forced to look at American culture from the outside." Living with Danish and other international students, learning to see differences among students from Communist countries, and sometimes avoiding standing out as an American all pushed the boundaries of Dan's personal identity. The Danes saw Americans as living to work and themselves as working to live. He found himself "endeavor[ing] to have it both ways."

Seeking to integrate work and life after his experiences abroad also highlighted a conviction that engineering is about service. During the Workshop, Dan said he has "an industry bias, if you want to call it that." He does not want to call it that. He maintains that important service work is accomplished through industry, and private companies are key vehicles for engineering work to contribute to the production and distribution of benefits. Arguing with a Trotskyist in Denmark reinforced his view that capitalism generates many inequalities but is better than any known alternative for producing and distributing benefits.

Dan's commitment to service also led him into other engineering activities. One was department and college-level administration, first at Arizona State and then back at Purdue as head of mechanical engineering. He also involved himself in national and international standards organizations. The engineering research discussed in standards, Dan observes, assumes deployment "in the developed or near-developed world" and gives little thought to the "developing world." It is important to Dan that he did not see this until after he became involved in international and global engineering education.

Nothing in a department headship forces an administrator to throw himself into building educational programs for students beyond the United States. But throw himself Dan did. He linked the new work to his established engineering identity by treating the challenges as design problems. Developing new programs was for him design work to meet the specifications of "cost neutrality," "no delay in graduation," and "development of a community" of support. His expectation at first was to support students seeking employment in industry. Successfully winning jobs in industry is an indicator of successful service. The experiences and identities he sought for students expanded when he was pushed by some who were "interested in serving humanity directly rather than through developing products or services via a commercial enterprise." Dan's view is that "all engineers actually help people," including those working exclusively in the commercial realm. All need expansiveness in both work and life.

An expansive life ultimately insures expansive work by revealing the limitations of narrow work. A global engineer is for Dan an expansive engineer—one who incorporates requirements for engineering problems "from multiple cultures," is capable of resisting "instinctive decisions… appropriate for their own culture," and actually making decisions "in a cross-cultural environment." The alternative is unsatisfying and maladaptive narrowness.

3. FITTING THE PRACTITIONER TO THE PERSON—MARGARET PINNELL

After completing her master's degree in mechanical engineering at the University of Dayton, Margie Pinnell worked for the organization in which she had done co-op work. She found her new position in technical program management wholly uninteresting. Equally important, she became acutely uncomfortable managing programs "developing war-related materials." As a permanent employee, Margie now belonged to the organization, and its international identity was about war. She found contributing to such work to be "in contradiction with my pacifist beliefs and values." She had already established an international identity for herself through pacifism.

Margie had never intended to involve herself in international engineering education. What she did want was "alignment" between engineering "activities and research" on the one side and "my core Catholic values" as a person on the other. Alignment was for her not just about minimizing conflict or incongruence among her identities. It was about redefining the boundaries of engineering on the job, expanding it into a vocation and merging it with life more generally.

Raised by a teacher mother and carpenter father, Margie initially pursued engineering because she was good in math and science and wanted financial security. The desire to link person and career through engineering did go beyond socioeconomic status, however. The University of Dayton was attractive because its Marianist Catholic identity made her feel at home and promised opportunities beyond her curriculum to "learn more about my faith." After leaving her job, the master's degree was something of an afterthought, the product of collecting graduate credits in engineering. The Ph.D. was different. It posed the new opportunity to add the identity of engineering professor. This pathway promised service through teaching, and with it the opportunity to have an impact on the lives of young people as "some of my engineering professors had on me."

Margie's description of "taking time off" to be a "stay-at-home mom" does not seem much like taking time off. During the ten-year period before beginning to teach, she gave birth to and raised three children, "worked part-time evenings towards my Ph.D., and even took on a job as a "quarter-time researcher." For her, "multitasking" meant juggling activities in a way that insured no identities inside or outside the home were ignored or allowed to drift into the background.[32] Margie highlights the extraordinary support she received from engineering colleagues during this period.[33]

While Margie calls attention to the attractions and tenure risks of diving into international service learning for engineering students, also notable was her insistence on making sure it was "sound pedagogy." Researching service learning extensively, she learned that both students and "community partners" have a "broad range of good and bad experiences with service-learning." The latter, in particular, "can sometimes feel exploited for the benefit of the student." Serving as faculty director for a student-founded group, she went to work figuring out how such projects can be "properly facilitated." She maintains that engineering students working over ten-to-sixteen week periods can genuinely "learn with [a] community about a specific technology." The experience can help students

---

[32]For an investigation of such multitasking as preparation for engineering work, see McLoughlin, "Success, Recruitment, and Retention of Academically Elite Women Students without STEM Backgrounds," 2009.

[33]Compare Pinnell's experience with the findings in Wasburn et al., "Mothers on the Tenure Track," 2009.

develop technical and non-technical skills; build self-confidence; enhance problem-solving skills, creativity and their ability to adapt to difficult situations; make connections between classes; develop racial and cultural sensitivity; enhance a commitment to civic responsibility; and increase ethical awareness and awareness of the impact of professional decisions on society and the environment.

All these, in her view, dramatically expand the practices of engineering work.

Margie took tenure risks because she wanted to be the kind of professor who was "able to provide the fuel a student needed to turn a spark into a flame—and to turn a dream into a reality." The desire has been to making sure engineering is integrated with passion. Margie's work in international service learning re-defines engineering as knowledge-laden practice that exudes passion from its core, i.e., is built on and engages personal commitments. Making war-related materials motivated the initial change in direction, but the real challenge has become transforming established engineering pathways that sharply separate the engineering practitioner from the person, such as what she found in technical program management.

4. REDEFINING THE ENGINEERING SPECIALTY—JOE MOOK

Joe Mook describes himself as having followed a "typical career" at a "large, comprehensive research-intensive doctoral university," the State University of New York at Buffalo. He "concentrated . . . on research and scholarship with my technical specialty areas . . . wrote research papers . . . was Principal Investigator . . . supervised graduate student research . . . taught engineering courses, participated in professional society, journals, and conferences . . . and performed occasional 'service' work on and off campus." In this description, personal identities needed no mention. These were sharply separated from his work identity.

The marginality of the personal began to change when, after becoming full professor, Joe had two experiences visiting technical universities in Germany. There the personal forced itself on the engineer. He "ate new foods, heard new sounds, saw new sights, [and] discussed new politics and culture." Routine life activities had the effect of generating "a completely new perspective," a "broadened perspective," that "permanently changed the way that I have processed world events and interpersonal relationships ever since."

Joe's dean evidently saw a change in him, for two years later he invited Joe to formalize international programs for the college as Assistant Dean of International Education (IE). What Joe found in this work became over the next several years what he remarkably describes as his "primary academic specialty focus." He built this focus by adding new identities himself through his position, albeit on a quarter-time appointment with no dedicated support staff.

Joe accepted he was, in part, an IE administrator but saw himself mostly as a faculty member focused on IE work. He was interested less in the administrative work of negotiating and managing partnerships, launching off-campus programs, recruiting incoming students, and engaging in strategic planning, and more interested in seeking outside funds, participating in professional organizations (including educational consortia), research and publications, organizing "meaningful, successful study-abroad programs for students," and successfully supporting and managing all dimensions of exchange relationships. He cooked up an especially ingenious plan for having outgoing undergraduate students generate tuition waivers for incoming graduate student research assistants. He also describes how important it was for him as a faculty member to have colleagues in his specialty, especially by participating in the Global Engineering Education Exchange program (see Lester Gerhardt's personal geography for more on the Global E3).

From one point of view, Joe's account is of a tenured engineering professor who decided in midcareer to redirect his interest away from nonlinear dynamics. But this shift was not simply to another technical specialty, a change that would likely have carried few implications for the relationship between his work and personal identities. In the actual shift he made, the personal dimensions of his identity moved from the background to the foreground, and the boundaries around specialty work in engineering became powerfully blurred. With this blurring came a compelling sense of career fulfillment.

5. COMING TO SEE THE PURELY TECHNICAL AS A BOX—ANU RAMASWAMI

Anu Ramaswami was "los[ing] much sleep" over the "ethical ramification" of her involvement with the "dynamics of donors, benefactors, beneficiaries, and their varying agendas" in a remote rural part of India. Having earned tenure in environmental engineering at the University of Colorado Denver, she had set out to escape isolation with theory and help produce the "real-world applications of engineering in developing community projects." She wanted to "have a positive impact on underserved communities." In part, she was returning to her country of birth and initial education where she had completed a degree in chemical engineering at a prestigious institute of technology. But it was not about connecting engineering to an essential Indian self or identity. It was about letting herself go and exceeding boundaries she had accepted in herself as an engineer.

In going to India, Anu was finally accepting a long-standing practice she characterizes as "stepping out of . . . boxes," especially technical boxes. In this case, it was about "creating energy and water resources for community use" in ways that would "resonate . . . with me intellectually, philosophically, and professionally." Resonance might happen if the work "preserved nature" and genuinely "connected supply-side technologies" with the "user sphere and the environment."

For Anu and the other two contributors who were born outside the United States (Bernd Widdig and Juan Lucena), adding a U.S. identity through immigration was not the life-changing experience that leaving the country was for most of the American-born participants, including me. The main attraction to Anu was increased freedom to step out of boxes, both as engineer and as woman. Drawing agency and self-confidence from the women in her family at least as much as from the men, she had spent much time as an undergraduate reflecting on how engineers and the outcomes of engineering actually interacted with and served (or not) the broad and diverse world of users. The U.S. stood for an opportunity to pursue broader questions in a more amorphously-defined field—environment engineering. Adding a new geographical identity was about expanding a work identity.

Humanitarian work in India and as a technical adviser extrapolated practices of collaboration she had developed prior to tenure. Modeling aquifers, examining the bioavailability of pollutants to microbes, and studying pollutant transfer in trees had required travel across disciplinary boundaries and attendant awareness of the limitations of her own knowledge. But these moves did not cause her to lose sleep. She was losing sleep in humanitarian work because she understood how little she knew about users, an awareness that dated back to the "ethnographical field methods" class she had taken in graduate school. Anu could not properly assist (i.e., unambiguously) without knowing who sets the agenda, whose knowledge counts, and who benefits (according to what measures). But awareness of her ignorance did not mean she had to retreat to narrowly-defined practices of technical problem solving. Engineers who were willing to step outside of technical boxes could find ways of making boxes bigger and connecting them with boxes built by others. Anu came to believe she could learn more about communities around Denver by interacting with them directly than she could learn about rural Indian communities she could engage only through complex mazes of mediating organizations, including both NGOs and layers of government.

Anu threw herself into practices of research and graduate education that pushed herself and her colleagues to find ways of tapping and engaging important knowledge that existed beyond their own. Such involves recognizing the limitations of one's own knowledge and identity. A course on defining and measuring sustainability, for example, requires students to administer surveys to "understand sustainability aspirations of the community." Learning about such aspirations and a range of other community complexities requires, among other things, a significant investment of time. Anu came to be an advocate for training engineering students in "community-partnered sustainability projects" at home before sending them off to other countries to partner, hopefully, with "elected government officials."

Such graduate education both inside and outside the United States counts, in her view, as global engineering education, for global education involves learning that "local actions have global impact on people and the planet, and vice-versa." Equally important is what the lead instructor learned about herself. She could both value and practice the quantitative rigor of technical boxes in engineering while also challenging their limitations.

6. QUESTIONING WHAT YOU DESIGN—LINDA PHILLIPS

Linda Phillips requires students in International Senior Design to spend forty hours as volunteer laborers on a construction site "in country," usually Bolivia. The students work alongside "local workers," in order to learn "local tools, materials, and techniques" in construction. She writes that, "[t]hrough my own industry experiences," she came to believe that "a sustainable and constructible design is generated only when an engineer understands the situation in the field." She wants students to add field identities that challenge and shake up as dramatically as possible their expected identities as construction engineers.

Perhaps most remarkable is that Linda is teaching senior design at all. She had "grown discontented" supervising construction work that "encourage[ed] the expansion of an already overbuilt environment." Responding to an invitation, she left behind "a job, a husband, a house" to go teach a Virginia Tech course for construction engineers on professional practice contracts and specifications. Linda became a construction engineer who was also an educator. Yet she found herself "in tears" asking, "How can I teach students to 'develop' more unneeded facilities?" It matters, she had come to believe, what construction engineers design.

Integrating the question of what to design into design education would prove far more difficult than Linda had anticipated. Moving to Michigan Tech, she served as an instructor teaching the professional practice course in civil and environmental engineering. Thinking it important to help students learn to see and critically analyze typical U.S. business practices in construction, she invoked another identity she had developed in recent years—participant in construction projects outside the United States. She had worked outside the U.S. both within industry and as a two-time volunteer on missionary projects with her mother. Linda began asking students to write specifications for projects that used "developing-world construction material and techniques."

Pressure from students and new ABET accreditation criteria made feasible a new type of capstone design experience. The practices she designed began with two weeks in-country learning construction techniques and meeting with "local project 'clients' and benefactors as well as with local design professionals and city officials." The latter gather design data that students formalize into "design-option feasibility studies," final reports with design recommendations, and draft construction contracts. The reports and contracts go to the clients.

Like Rick Vaz, Linda relies heavily on the image of comfort zone, believing that students can best learn how to question their routine construction practices if they are deprived of their "culture, language, daily routine, food, pastimes," and accept "tight schedules, an early curfew, limited cell phone and internet use, roosters crowing all night, doing laundry by hand, no TV, and the milk lady squawking through a bullhorn by dawn." The physical labor on job sites also tests them physically in new ways and they "quickly appreciate the construction workers' strength and stamina as well as their wisdom." The ultimate payoff for Linda is knowing she is helping prepare construction engineers who do not simply accept, as given, projects to pave over green spaces and who tend to insist on asking if the project is constructible.

Working on a construction site, however, is not in and of itself design work. Design outcomes that rely on seemingly primitive construction practices may not seem compelling to engineering science faculty who are focused on pushing the theoretical boundaries of their specialties. Linda's personal geography is a detailed account of a struggle to enable engineering design education to help students learn to question what they design.

7. SERVING AS RESOURCES FOR MORE THAN A COUNTRY—LES GERHARDT

The interview at Rensselaer Polytechnic Institute was partly an opportunity to visit friends. Les Gerhardt was by any reasonable standard enormously successful as an electrical engineer at Bell Aerospace. While working on the visual simulation of space flight, he had "risen rapidly, technically and managerially," winning the Bell Outstanding Management Award. But he "relinquished a solid position" and decided to "try out academe" in part because he had found teaching at a local community college in the evenings to be both interesting and important.

Working at Bell and gaining an identity as a Bell engineer had transformed his understanding of engineering from something people who are good at math and science do to the intelligent use of resources for planetary benefits. "It was then," Les writes, "I experienced a real awareness that the science and engineering workforce needed to be regarded as more than a national resource." His focus as an engineer was expanding beyond advancing visual simulation to helping design the engineering workforce. And with this expanded geographical focus and new identity came a new type of obligation.

The interactions in this Workshop made it clear that the contents and practices of administrative work in academic arenas are poorly understood and little analyzed beyond the narrow academic arena of higher education administration and leadership. Universities in particular are so dominated by images of discovery and creation that they place everything else in subordinate roles of either diffusion or support. Les wants readers to know his contributions to research, graduate advising, and consulting in digital signal processing continue to this day. What really stand out in Les's career trajectory are his many commitments to time-consuming administrative identities and activities both on and off campus. He has worked on campus as department chair, center director, associate dean, dean, and vice provost, among many other things. Administrative work clearly provided deeply-satisfying knowledge about what is involved in building the engineering workforce, and hence what might necessarily be involved in re-defining it. Much is embedded in his assertion that academic administrative work "permitted me to reach out and help many faculty and the university as a whole." Reaching out from elite administrative positions came with the authority and power to formulate and enact visions. At whatever level he was operating, Les's attention was to the whole.

Les's off-campus work as a NATO delegate, international adviser for the American Society for Engineering Education, and most powerfully as co-founder and longtime leader of the Global Engineering Education Exchange (Global E3) Program, all illustrate how much his new identities, and hence his new purview, lay beyond the country. Perhaps it had been working in the defense industry before the image of economic competitiveness scaled up to dominance that both grounded a sense of the "futility of war" and kept advancing private-sector production for mass consumption from becoming the fundamental career activity and goal.

Les himself highlights the crucial knowledge he gained through international travel. It was by being "exposed . . . to many different cultures from the 'inside'" that made him aware "of the difficulties faced by the citizenry of many countries." For Les, extensive international travel brought with it new personal identities, including connectedness to others through cooperation and critical awareness of the privileges he enjoys. These personal experiences were crucial in supporting a revised image of the larger service responsibilities of engineering. "This was not TV," he asserts in acknowledging the privileges of wealth.

Just as engineering practice carries normative commitments to broader social projects, engineering education carries an obligation, Les asserts, to produce competent practitioners aware of those commitments. As he approaches the end of his career, Les is working to insure that awareness of the possibilities of cooperation and the responsibilities of privilege emerge from required rather than optional curricular experiences in engineering. Globalizing engineering is necessarily about remapping the identities of engineers beyond the boundaries of the country.

8. JUXTAPOSING THE HUMANITARIAN WITH THE COMPETITIVE—ALAN PARKINSON

Alan Parkinson had "some sleepless nights" when he realized he was one of two final candidates for the engineering deanship at Brigham Young University (BYU). He "accepted with some trepidation" because he knew the vision he had presented would be difficult to implement. Alan had been looking at the College of Engineering through the twin lenses of a specialist in design optimization and holder of a recently-awarded MBA (master's in business administration). He judged his responsibility to be enhancing the competitiveness of a predominantly undergraduate institution with a faith-based identity. His approach was to identify constraints and resources.

One of the greatest resources lay in the fact that 75% of the engineering students who had previously served as missionaries (half the women and a significant majority of the men) spoke a second language. Alan's vision for the college thus included "global awareness" as a strategic outcome. This strategic outcome was authorized by new ABET criteria and prospective employers. Successfully realizing it would make the engineering college at BYU better known and help it remain competitive in attracting quality prospective students and then placing them in quality positions.

Like Rick Vaz, Dan Hirleman, and Margie Pinnell, Alan returned to serve at the institution within which he had earned his undergraduate degree and passed into full adulthood. For each of these four, educational program development and administration is also about enhancing the visibility and status of his/her alma mater, giving it special personal as well as professional significance (Gayle Elliott, below, also teaches at her alma mater, earning her first degree working part-time as an adult). For Margie and Alan, the connection has additional personal significance because they work at faith-based institutions. There maturation of the person is judged to be as much about the solidification of faith as it is about age and educational advancement. Alan's case stands out even in relation to Margie's because his institution stands at the center of the religious activities that support it. As he describes it, the Church of Jesus Christ of Latter-Day Saints (LDS Church) invests "hundreds of millions of dollars per year" in BYU, amounting to "70% of the operating expenses." It also has more than "50,000 young adults" serving at any moment "in hundreds of missions across the globe." Both the Church and the University are based in Salt Lake City, Utah.

A key expectation for a BYU alumnus teaching at BYU is that one's work and personal identities are wholly in alignment. This is a given. It is also instructive that the personal identity of an LDS Church member is often international in content. Alan declined admission to MIT for his undergraduate degree because he "had been planning on serving as a missionary for a number of years." Mission work is frequently international. To prepare students for international positions in industry or anywhere else is thus wholly in line with one's dual, parallel identities as faculty member and member of the LDS Church. It would be a surprise if the engineering practitioner and the person were not wholly merged in this way.

The connection between academic and religious identities extends to BYU academic units. The technical output of an engineering department at BYU is not diminished by a mission not only "to provide a world-class education to undergraduates and graduate students" but also to do so "in an atmosphere enlightened by the principles of the gospel." The gaining of spiritual knowledge "by faith" is a stable, expected, and unproblematic feature of the department's identity. It is a given that a key responsibility of engineers, like everyone else at the institution, is to be "an influence for good in the world." What is variable across academic units is the gaining of "secular . . . knowledge 'by study,'" which places graduates along differing career trajectories. The College of Engineering at BYU is not only in competition with other engineering institutions for students. It must also retain its position within the LDS Church as the Center for the production of Mormon engineers.

Alan's trajectory is especially remarkable in that he completed the executive MBA program at night after finishing a term as chair of his department. He was worrying that spending the rest of his career writing proposals, doing research, and then starting over again would become "stale."[34] He was even considering leaving the academic world entirely. Gaining new appreciation for such areas as "strategy, organizational change, human resource management, leadership, operations and product development," in fact, accelerated his thinking about American engineering in strategic terms.

Alan's personal geography thus offers a case of an American-born engineering educator for whom adding an identity outside the country prior to developing programs in international and global engineering education did not produce a jarring incongruence. Arriving in Japan for his missionary assignment had indeed felt "as if I had been transported to another planet." And he had returned home "with great respect for the culture and accomplishments of the Japanese people." Yet his identity as LDS Church missionary had been enhanced rather than fundamentally challenged. And the amazingly-rapid scale-up of international and global engineering education at BYU that he led and describes was also a strategic extrapolation of engineering education for national competitiveness. Rapid economic development in China was, as Alan quotes a colleague, "a train on track coming right at us . . . . If we don't move we will get run over."

It was the experiences in developing and running international and global engineering education programs themselves that challenged and ultimately expanded Alan's understanding of engineering service to include "humanitarian" goals and activities. He was impressed that his director of the Mexico program had been affected deeply by first-hand experiences of "extreme poverty,"

---

[34]This quote is from Alan's first draft, included with permission.

producing a desire to "make a difference." Watching the development of a small-scale facility for converting coconut oil to bio-diesel fuel in Tonga made him refocus on the University's motto, "Enter to learn, go forth to serve." He found that prospective donors to the College were especially receptive to "humanitarian study abroad programs." Feeling pressure from department heads to place greater emphasis on research led him not to consider reducing international and global engineering education but to begin considering approaches to increasing the "efficiency" of learning and teaching. Without giving up on economic competitiveness as a goal for both the country and the individual student, he found himself attending increasingly to the importance of recognizing and understanding privilege. Students who participate in humanitarian projects, both "develop an understanding of themselves as members of the world community" and are "often humbled by what they enjoy as citizens of the United States."

9. REDEFINING MEASURES OF PERFORMANCE—JAMES MIHELCIC

"You know, we should do that," said Jim Mihelcic to a senior colleague over coffee. He was referring to a new master's-level international program in forestry at Michigan Tech. The program merged education and research for the degree with two years of service in the Peace Corps. Jim did not follow the colleague's counsel to stay focused on his research and advising in environmental engineering, especially methods and practices of remediation. But then he had not followed his father's advice to stay away from environmental engineering in favor of mechanical or chemical engineering, which would have prepared him better for a second degree in business or law. Jim reports he has tended to follow pathways that "integrated my heart and personal convictions with the structure of my engineering brain." This time he would be confronting "university measures of performance." He would find that "[s]tanding up to colleagues and administrators who question or block your ideas to innovate engineering education is . . . not easy to do." Jim had not physically left the country, but he had added a new international identity nonetheless.

Jim got only one student in each of the first two years of his new program. He had successfully figured out how to incorporate into an existing graduate program a semester of Peace Corps training, at least six semesters of Peace Corps service, and "overseas research." The most significant immediate implication was the need to collaborate with faculty beyond his department. Early on, it also involved accepting a two-credit overload in graduate teaching—so he could make instruction in appropriate technology "rigorous and worthy of graduate credit."

When interest expanded to a half-dozen students by year three and then as many as a dozen a couple of years later, Jim had to give up his graduate course in environmental remediation. Now fully accepting the adaptation of his work identity to a personal mission, he had decided to focus on "proactive approaches related to the prevention of pollution." Intellectually, this meant shifting research and educational practices to embrace the emerging images of "sustainability." Educationally, he was also responding to increased numbers of students "expressing stronger desire for programs related to environmental degradation, climate change, poverty, justice, renewable energy, and serving others."

An important learning experience for Jim was finding out that integrating new practices into existing curricular structures would also require dramatic alterations in his daily practices of work and life. "I changed my daily routine," he says for example, "to be responsive to students' requests." Changes included responding to email requests within twenty-four hours, tracking down literature on campus, or "quickly packaging some research supplies for the one-month journey to a student's site." They included building new types of research groups, such as one in Bolivia in which nearly half the thirteen students were Bolivian, and another in which U.S. doctoral students trained with undergraduate U.S. and Bolivian students. He found himself exploring ways to define research topics "that would be acceptable to a graduate committee and outside peer reviewers." He had to figure out how to "move a research publication through peer review when perhaps the reviewers are skeptical of the quality of the research coming from a village in Africa," and to persuade promotion and tenure committees of the value of new bodies of literature that are "more interdisciplinary and perhaps have a more applied audience of nongovernmental organizations." Making these changes meant sending students to courses in anthropology and geography. They meant devoting enormous time and energy to writing a new field guide and textbook.

In parallel with perhaps all the other contributors to this volume, Jim goes to work each day trying to "merge [personal] conviction and values with my profession." Also like many, merging person and practitioner redirects the outcomes of emerging work in new directions. For Jim, the resulting changes meant regularly putting himself at odds with faculty at one university who "stood outside in their conventional educational and research world." Conventionality did not make them wrong or bad, Jim insists. The daily experience of sharp contrast did, however, generate in him a longing for a scholarly environment more congruent with his new identity. He was greatly relieved to find one in which the larger objectives of his work fit the explicitly-defined strategic mission of the university. And he finds affirmation in responses from students who, like one, indicate "that along with gaining valuable engineering skills, he also learned what it was like to put engineering into practice while taking into consideration the social, economic and environmental limitations of the developing world."

REDEFINING ENGINEERING: FIVE NON-ENGINEERS AND TWO HYBRIDS

For the five non-engineers and two hybrids contributing to this volume, adding identities beyond the borders of home countries fueled ambitions and practices to help engineering students better recognize and analyze the broader commitments in their work.

1. ENGAGING PRESENT REALITIES—BERND WIDDIG

"I soon asked myself how I could connect my work and interests with the large engineering and science community at MIT." In asking this question, Bernd Widdig admitted a possibility into his daily work and professional identity that his senior colleagues in the Foreign Language and

Literature Section evidently abhorred. Bernd built a successful international program that sends MIT students to internships with companies in Germany. Seeking this connection and identity across the MIT campus with clear vigor and insight cost Bernd his tenure and promotion, an experience he courageously discloses and explains.

The effort also propelled him forward into an administrative path that he characterizes as "deeply satisfying." Recognizing the depth of this satisfaction, readily apparent to all who know him, calls attention to the ways administrative work can provide opportunities for redefining the contents and boundaries of educational practices. In this case, following the pathway leads to rethinking the relationship between engineering and the humanities. They do not have to be radically distinct intellectual arenas with an unbridgeable gulf between them.

Bernd had received his graduate training in German studies at Stanford. There he sought out and endured the practices of extended experiential learning that he aspires to provide for his students. Bernd is a native German who had studied economics, political science, and German literature at the University of Bonn. In contrast with the dominant learning practices among scholars in foreign language and literature, his main interest was not in using language learning as a pathway "to read increasingly complex literary texts." He did not want to pursue a career pathway into the isolated study of "the established literary and cultural canon of a country," in this case Germany. Rather he was much more interested in "the present cultural reality in Germany" (and elsewhere).

This interest in the present is what grounded his acceptance of an invitation to join the German studies program at Stanford. While "area studies" fields still grant highest status to the literary canon, they are also interested in the "complexities of a political, anthropological, and economic everyday culture." Bernd's interest in connecting the literary to the everyday was illustrated by the focus of his second book, *Culture and Inflation in Weimar Germany*. It has also been illustrated by his openness to allowing his personal identity to evolve through life experiences. He became, in a way, a "sunny Californian."

Bernd insists that to become an "educated person" one has to be able to understand and reflect intelligently on the "messiness of real life" in the present. Organizing internships in German corporations for MIT students not only reversed the decline of enrollments in German language and literature classes, which was good for the Foreign Language and Literature Section, it also enabled Bernd to acquire a valuable, contemporary identity: engineering educator. Hearing from students on internships about "an exciting research project they were working on side by side with young Germans" or "listening to them about their daily encounters with German everyday culture" was for him crucial. It "gave me a sense I was making a difference in their lives."

In building MIT-Germany, Bernd was leaving the "religious order" of foreign languages and literatures, for he was adding something new to the dominant identity of research professor. To his colleagues, such work carried the danger of precipitating a decline into a "service department" and the much-feared identity of an "in-house Berlitz school without much academic and scholarly standing." To Bernd, the work was an attractive administrative and scholarly experiment in the "applied humanities," which promised to transform the unbridgeable gulf into practices of "creative dialog." Could the humanities and engineering come together in creative dialog even while continuing to serve distinct societal projects, all in the interest of achieving education for the present? For Bernd, questions such as this one motivated new kinds of work to produce educated engineers for the commercial realm. They are now motivation for enhancing the technological literacy of Boston College students studying abroad.

2. CONTENT KNOWLEDGE AND LANGUAGE ARE CONNECTED—MICHAEL NUGENT

When Mike Nugent entered a Ph.D. program to study comparative higher education, it was attractive in part, because he would not have to choose again between an interest in language education and a so-called "content" area. As an undergraduate student, he felt forced to choose between degrees in music and in foreign language and literature. Despite the fact he had left high school early to begin jazz studies in college, a trip to Europe at age thirteen had changed him by adding an indelible fascination with languages. Indeed, before ultimately completing a degree in German language and literature, he had spent six months in formal language training in Germany and had also completed an intensive language course in Swedish. A Ph.D. in comparative higher education would allow him to link what was by then a defining interest in language education to an emergent interest in higher education policy. This connection would help him find positions constructing educational programs that would not require students to choose between language and content proficiency.

By the time he began doctoral studies, Mike's experiences in language education on both sides of the desk were extensive. He had spent a year in Hamburg as a student of German language and literature (*Germanistik*), a summer in an intensive French course in Tours, and a year learning Spanish in Madrid. For several years, he had earned income teaching English to professionals in Europe and to community college and international students in the United States. He had received advanced training in language acquisition through a summer fellowship program at Harvard. The desire to build an identity in higher education policy had grown with direct experiences of educational practices in Germany, France, and Spain. This desire was also traceable to the shock and awe he had experienced as a thirteen-year-old traveling across Europe. Directly encountering a myriad of linguistic differences had made him realize he lives in a world characterized by multitudes of differences.

Mike's personal geography offers another case for expanding attention to the opportunities for redefining intellectual boundaries through administrative practices and work. Before becoming an administrator of educational programs, Mike had first-hand experience of the pleasures of making a difference in a classroom. The analysis, implementation, and formulation of policy grew as an object of interest because it offered the possibility of formative influence at much higher scale. After working for a state higher education organization and then helping set up a national body for assuring quality in higher education, Mike moved to FIPSE, the U.S. Department of Education Fund for the Improvement of Postsecondary Education. Given the identities he had acquired outside the United

States, Mike went to work managing and building international consortia programs. These included especially a European Union-U.S. program, a North American program, and a U.S.-Brazil program.

FIPSE proved attractive to Mike because its international programs employed what program officers came to call "meaningful time abroad." Time abroad for students became meaningful when it went beyond menus of elective courses to include learning experiences in the student's primary field of study. It was in this context that Mike began to encounter projects that distinguished themselves from "international education" (decoded as "study abroad") by moving to "global education." The latter was both more variable and frequently enacted an amorphous image of preparing students for a "global workforce." FIPSE work also involved directly confronting the barriers among participants in international consortia that prevented student experiences from scaling up in systematic and smooth ways. The barriers involved everything from the absence of workable credit transfer practices (especially when the concept of awarding credits did not exist in the first place) to inequities in flows of students across the borders of countries.

Despite the dedicated work of Mike and his colleagues, "the majority of [FIPSE] projects tended to avoid the language issue." This meant he was now complicit in forcing students to choose between language education and content knowledge, even when his experiences teaching English to professionals had made it obvious the two are necessarily connected in the everyday lives of working professionals. The United States is the only country in the world, Mike points out, in which learning a second language is not a career and life expectation.

Mike now finally feels fulfilled, indeed living his calling, because he leads a large program that demands a deep, abiding linkage between content learning and language learning to the highest levels of competency. He now has the ability to insist students not have to choose between the two. To him, this means they are now more free to discover that they (like him) live amidst a multitude of differences, which they must learn to navigate.

3. EXAMINING ENGINEERING LIFE THROUGH LANGUAGE AND LITERATURE—PHILLIP MCKNIGHT

"Do we want to produce engineers—or other graduates for that matter," asks Phil McKnight of Georgia Tech, "who will work quietly on their projects and in their labs producing commercially viable products without regard to their impact on the environment or economic structures?" Or, he continues, "do we want to produce an intelligentsia that will understand how to employ technology, science, and the humanities in ways that have a principled, positive impact on society and the world in which we live?" Phil had earned full professor in German language and literature at the University of Kentucky, in part, through scholarship about Christoph Hein, one of the most famous writers of the former German Democratic Republic (GDR). Following Hein's writings and life, especially during seven summers in residence in the GDR, had shown Phil that literature really can be "connected to life." Literature, as "intellectual creativity," could "form a matrix of aesthetics, historical analysis, . . . and political events" to show "How people are affected by different aspects and implementations of public policy." The question of literature and education (by extension) became for Phil the question: "How can intelligence function to influence policy?"

When Phil was an M.A. student in the late 1960s, he organized a symposium on the question "Why do we study Germanics?" He had been ruminating on the story of Faust, who "rejects his isolated and frustrating life in a university research lab" to answer the call to "flee out into the wide world." Pointing out that perhaps 30% of German literary scholarship at the time was devoted to Goethe, author of the 18th century play, Phil describes wondering "[W]hy were we not taking his advice [and leaving too]?" This developing interest in the connection between literature and life spurred completion of an advanced degree, a "ten-year hiatus" from academics "due to financial necessities associated with single parenting," longtime work in a motorcycle shop, and eventual enrollment in a summer program "designed to transform Ph.D.'s in the humanities and provide them with tools to switch careers." Phil carried this question as he used the mediating identity of a journalist to build the relations he thought he needed to become a business leader who supervised the development of a free-trade zone in Colorado. And he carried the question during nineteen years as a faculty member and eventual Chair of the German Department at Kentucky.

Characterizing himself as an "incurable dreamer," Phil undertook forays into the world of business not because of a principled commitment to advance capitalism, as Dan Hirleman and Alan Parkinson exhibit. Indeed, Phil opens his personal geography by detailing inequities produced by business and engineering practices of deforestation. Rather, he turned to business as a critical participant, as someone who was resisting the retreat of language and literature studies into the abstract ("*l'art pour l'art*"). In a question that suggests he is referring not only to deforestation but also to the negative effects of any type of engineering or business practice, Phil asks: "Would any of this have been done differently had experts in culture, language, and society been able to exert their influence in a credible manner, or by integrating these views collaboratively with those held by business, engineering, technology, and [the] policy sector?"

Phil insists that language learning must encompass "not only culture and literature but science, business, technology, politics, and international relations as well." The practices of language learning should not be organized to lead simply to esoteric high culture on the one side or professional skills on the other. Language learning should be about the enhanced understanding of self and appreciation for others. It should be about adding identities that challenge what is there.

It was important to Phil that Wayne Clough, the Georgia Tech president [head of the Smithsonian Institution at this writing], both understood "the lack of communication between technology and the humanities" and was committed to educational innovations in which each is transformed by the other. This sense of hope and optimism grounds Phil's dedicated efforts to help build Georgia Tech's International Plan, an ambitious enterprise designed to insure that at least 50% of its undergraduates complete an experience outside the United States. Phil thus contributes willingly to a "pragmatic engineering approach" to helping engineering students build "global competency" because he "knew that the immersion experience abroad would bring about" his real objective— "changes in the students' mentality." The "acculturation taking place" while students are abroad, he believes, offer perhaps the best chance for overcoming "the 'ugly American' syndrome." By this he means "produc[ing] in such individuals high ethical standards and a great appreciation for developing mutually beneficial business deals that would not leave the global impression that the U.S. was ruthlessly exploiting other countries for profit. . . ."

4. CHALLENGING BASIC VALUES AND POLITICAL ASSUMPTIONS—JOHN GRANDIN

Referring especially to engineering students, John Grandin writes that "Americans will nod politely at the idea of learning languages to broaden one's horizon." The intrinsic interest is just not there. They "will show up voluntarily in the classroom" only when "they are convinced that it will impact their lives and strengthen their position for career success." The "American public," he continues, is "years, if not decades, away" from recognizing the "wisdom" of international education. John is thus insistent in the view, like Bernd Widdig, Mike Nugent, and Phil McKnight, that "language learning must be rooted in many subject areas." John also asserts it must be tied to "the phenomenon of internationalization, global awareness and the need to keep the nation informed and competitive." Accordingly, John's International Engineering Program at the University of Rhode Island (URI) treats as full "partners" the corporate donors that have underwritten the program. Yet he is driven not by a consultant's contractual commitment to help engineering students "pad resumes," as one Workshop participant put it, or improve corporate balance sheets. He is driven by the memory of "finding myself and my world view challenged . . . on a daily basis" as a college student living with a family in Germany. He has taken on the responsibility on is to make sure as many students as possible have the opportunity to confront similar challenges.

John was sufficiently discouraged about his work as a German professor that he considered leaving to become a building contractor. He even joined his brother in building a house on speculation. John had arrived at URI after completing a Ph.D. in German literature at the University of Michigan, his dissertation on Franz Kafka. Learning literary theory and analysis had been remedial work, for his interest was less in literature and high culture than in languages. Having taught German for two years at Union College, John had discovered he could impart to students "not only knowledge of the language, but the importance of understanding the perspectives of people from nations other than our own." The experience was both "powerful and formative" and became key to his understanding of how to "make a difference."

John had been surprised he enjoyed German class while a first-year student at Kalamazoo College, interested in philosophy but not following his father's footsteps into the clergy. After a second year of study, he accepted an offer to spend the summer of 1960 in an intensive language course at the University of Bonn, living with a host family that had fled Eastern Europe and needed additional income. He was shocked to realize just how much he and his American friends took their lives for granted and were "totally unaware of the life experiences and priorities of people in other parts of the world, or of their expectations and hopes for us." The experience of finding "many of my own basic values and political assumptions challenged" became a spur to another year abroad, deeper understanding of other perspectives, and ultimately to a career "figure[ing] out ways to get students over to Germany for experiences such as I myself had known."

John's account describes the painstaking work he has put in for over two decades to build a program straddling the College of Arts and Science and College of Engineering for getting engineering students "over there." He details the logistical struggles and identity work involved in building teaching and research practices in applied language learning into a five-year double major in engineering and languages, an exchange program and dual graduate degrees with the Technical University of Braunschweig, a Center for International Engineering Education from remodeled fraternity houses, new programs in French, Spanish, and Chinese, and a national annual Colloquium on International Engineering Education. Each addition further expanded the number of students who will find their world views as both engineers and people challenged on a daily basis.

5. OPENING THE EYES BY ENGAGING THE FOREIGN—GAYLE ELLIOTT

The coordinator for the Japan portion of the International Engineering Program at the University of Cincinnati had left the University while the first group of students was working abroad. Gayle Elliott, coordinator for the Germany portion, began working with students on both sides. Taking this step added a new work practice—working directly with employers of the international co-op students. Students in the Germany co-op program had been receiving their job placements from an outside agency. For Gayle, to accept a new job identity with new responsibilities was in itself not especially new. Since she had first gone directly to work after high school, she had eagerly added new identities and work practices that showed evidence and promise of "moving up." "I derive satisfaction," she writes, "from knowing I do something well." This last step was especially difficult, however, in that she became more directly involved in structuring the experiences of engineering students participating in co-op job placements abroad. Taking the step linked a long-standing practice of making things work to the new practice helping "open the eyes" of students to the benefits and pleasures of engaging the foreign. To engage the foreign was a desire that had taken her to the Soviet Union after high school graduation.

Neither parent had attended college. Gayle's father had worked as a salvage diver for the Navy, and her mother worked in an office. A class on Russian history and culture had piqued an interest in a first-hand experience of the exotic. The experience did not disappoint. Getting off the plane, "everything was different." Not surprisingly, she writes, "we were not permitted to go anywhere alone," talk to people, or give them gifts. "It was communist," she asserts, emphasizing the foreignness of the encounters. Seeking foreignness did not, however, become her vocation. It became her vacation. A honeymoon ten years later to England, France, Italy, and Greece "whetted my appetite for travel." Gayle and her husband traveled internationally whenever possible, in part because it allowed movement past the routines of "nine-to-five." It also added identities, affording the knowledge that, for an increasing number of countries, "we'd been there."

Gayle had not seen the connection coming between a work identity and a personal identity when she moved from a program coordinator's position at the medical school to assisting an associate dean in engineering. It was about greater potential for upward mobility. Nor had she seen it coming when she had been pursuing an undergraduate degree in business management, both enjoying the challenge of learning and "obsessive" about studying and mastering content. She even did not see it coming when she went on to earn a master's degree in health planning and administration, a credential that could have carried her back to the medical school.

Gayle only saw the connection between the practitioner and the person who had long engaged the foreign when, as she puts it, "realized I enjoyed working with employers" in other countries. She saw it when she realized she "loved working with the students" who were planning to work in other countries. Those who accepted the challenge of learning a second language and then setting off abroad "came to me as immature freshmen, left to go abroad with excitement, and returned home as mature, changed adults, with tremendous self-confidence and the belief they could succeed anywhere." This she found exciting. Gayle was experiencing a benefit in gaining more responsibility in the International Program that was not primarily about upward organizational mobility and that no previous positions had provided. She was helping students "open their eyes" just as hers had been opened by a high school teacher and a unique travel experience years earlier.

Much of Gayle's account traces the difficulties in locating international experiences in the University's organization of roles and practices in cooperative education. To hang onto the recently-added identity of educator-who-opens-students-eyes-by-helping-them-confront-and-engage-the-foreign, she had to accept changing organizational locations, fluctuating budgets, and escalating responsibilities. Indeed, she concludes by reporting difficulties she has experienced after the international engineering program was made part of the University's signature co-operative education unit, the Division of Professional Practice. With that change, however, came an addition to her identity that neither she nor her parents could have imagined when she returned from the Soviet Union—she had become a tenure-track professor with a permanent responsibility to "do it again for another student," and then another and another.

6. HIGHLIGHTING AND TRANSFORMING THE POLITICS IN ENGINEERING—JUAN LUCENA

"Over the past five years," Juan Lucena reports, "I have found myself traveling with, consulting for, and visiting deans of influential schools of engineering in Brazil, Colombia, and Mexico." He has been "helping them figure out how to align themselves with opportunities for engineering education reform coming out of the U.S. and Europe." At the same time, he "confer[s] with U.S. engineering education reformers "who are trying to figure out where and how in Latin America to implement new initiatives for reform." He experiences ambiguity, he says, acting "both as a researcher/participant observer and as an activist who wants understanding and respect for the marked regional, ethnic and institutional differences among and with countries and their engineers." Juan's struggle is to make visible the "power relations" and politics in engineering in order to help "committed people trying to construct alternatives to engineering for military and free-market development." In the process, he is following in a way the footsteps of his father who as a Colombian political leader two decades earlier was trying to figure out what the decline of the Cold War meant for Latin America, especially Colombia.

Juan had bucked family tradition on his father's side by eschewing law and medicine for education and an identity in aeronautical engineering. "Three generations of Lucena lawyers and doctors had brought much prestige, power, and wealth to the family," Juan reports, but he could not follow suit. Despite his interest in aeronautics and high-performance engines, he found himself unwilling to contribute to what he has seen as Cold War or post-Cold War "militarism." He also resisted contributing to the 1980s spread of free-market ideology across Latin America. "[W]hat could I do," he wondered "that was both acceptable to family and appropriate to country?" While pursuing his engineering degree at Rensselaer Polytechnic Institute, reading and instruction in a course in interdisciplinary science and technology studies (STS) led to the conclusion that "engineering is always political." This finding initiated a longtime search for an alternative politics of engineering and technology. The search took him first to masters-level study in STS examining the possibilities of "appropriate technology," "intermediate technology," and "barefoot economics." Disappointed in the appropriate technology movement, he turned to government, completing a graduate internship that introduced him to research in engineering ethics as well as emerging initiatives to expand the pipeline for women and underrepresented minorities in engineering education and fund new arenas of engineering research and development. There he found emergent political arenas that seemed to place engineering at the center. Wanting to develop the theoretical sophistication for examining engineering and society at the same time, he began pursuing an STS Ph.D. at Virginia Tech.

Much of Juan's account describes his struggles transporting the Engineering Cultures course from Virginia Tech to Embry-Riddle Aeronautical University and Colorado School of Mines, where he encountered substantial resistance from disciplinary colleagues in the humanities and social sciences. As Juan explains his version of the course, it involves examining the expectations on engineers' service to society. A key strategy for him was to develop extended case studies for students to explore that were appropriate to the institution. At Embry-Riddle, the focus was on following the supply chain in the aerospace industry (which grounded an NSF Career Grant on engineering and the globalization of aerospace). At Colorado School of Mines, the emphasis shifted to extractive industries. An important innovation was to develop a wholly new version examining engineering cultures in Brazil, Colombia, and Mexico, enabling students to see each as "people with a complex history, respectable engineering institutions, [and] capable of forging their own history and images of progress." Taking this step propelled Juan to resist the resistance of colleagues and join with others in developing still more courses on humanitarian engineering and sustainable community development.

Teaching these courses at the same time that he added the personal identity of American citizen provoked him "to re-discover my Colombian self" and strengthened his resolve to challenge "the uncritical posture" of engineering students. Is the "expectation . . . of service to their societies as engineers coupled with desires to help, often motivated by religious missionary views all [students] need," Juan asks, "to try to solve other people's humanitarian and development problems?" His answer is a definite "No." His ongoing work is to persuade engineering students and colleagues to agree by helping them see and analyze the politics of engineering.

7. ASSESSING ONE'S KNOWLEDGE AND COMMITMENTS IN RELATION TO OTHERS—GARY DOWNEY

Gary Downey was stuck. Wanting to study engineers and engineering, he could not figure out how to approach them with the analytical tools he was gaining as a doctoral student in cultural anthropology during the 1970s. He had witnessed too much diversity among engineers to be able to study them as members of a distinctive culture with unique systems of shared symbols and meanings. Gary had made his way into cultural anthropology to gain the ability to analyze intelligently and participate effectively in conflicts over the outcomes of engineering work, especially controversial new technologies such as nuclear power. This desire had grown while he was an undergraduate student in mechanical engineering and found himself feeling "confined by the requirements of my curriculum." The specific identity he actually gained as an anthropologist would prove to be something he would have to overcome as much as perform or enact. The principal means would be helping engineering students learn how to analyze and critically assess their own developing work identities and expected future trajectories.

For Gary, becoming an engineer was a family expectation, which made adding the identity of anthropologist a sharp deviation. Dreaming in Spanish in high school had spurred him to seek a B.A. in Social Relations in addition to a degree in engineering, and then to take the "flying leap" into cultural anthropology. Gary prefaces his account of "critical participation" among engineering students with a description of what he calls his "two conservatisms." One conservatism involved simply accepting dominant practices, in his case practices of engineering teaching and learning designed to lead him onto upwardly mobile pathways in private industry. The other was the fantasy of the critic, believing he could find a stance or practice that would be located outside or somehow beyond dominant practices in the engineering production and use of new technologies but still somehow engage and change them. Gary rebounded from the latter into a possible career in science and technology policy but was dismayed to conclude such would require leaving engineers and engineering behind. He decided to try on the identity of an academic in the emerging field of Science and Technology Studies (STS).

Much of Gary's personal geography traces his efforts to participate critically in the formation of engineers and engineering identities. Three different types of courses he developed were ostensibly "for engineers," but their main emphases on technology prevented them from fully engaging the developing identities of engineering students enrolled in them. One type involved making visible assumptions about technology as an external force, a second examined relations between technological and socioeconomic hierarchies in the world, and a third made visible the nontechnical dimensions of technological decision making.

It was a fourth type of course that became definitive, beginning with the *Development of the American Engineer* he first taught in 1983. This type stood out because it focused explicitly on the "emergent identities of contemporary engineers in the United States." When Gary gained permanent approval for the course in 1988 as *The Technologist in Society*, its focus was particularly on what he called the "corporate technologist" in the United States. It examined such things as the emergence of distinct engineering fields, alternative approaches to engineering education and training, abortive attempts by engineers to form unions, ambiguities in engineering professionalism, and the gender expectations confronting engineering students. A key move in the 1980s was to integrate lecture material on engineers in other countries, especially Japan, which was then viewed by many in the United States as posing a threat to the country, much as China and India are often viewed today. The success of these initiatives fueled a seven-year effort to gain sufficient historical knowledge to transform *The Technologist in Society* into the new course *Engineering Cultures*. As Gary explains, he made the final move after developing "location, knowledge, and desire" as practices of critical self-analysis students could "take back" with them to their engineering curricula.

For Gary, developing a classroom-based effort in international and global engineering education has thus not been, in the first instance, about enhancing the competitiveness of engineers and the country, producing global citizens, improving the sustainability of life on the planet, or stopping the complicity of engineers in the exploitation of others elsewhere. It has rather been about helping students redefine their identities and work as engineers by enabling them to recognize and critically assess their own commitments to broader social projects. Such, he argues, is a necessary prerequisite to both analyzing dominant practices of engineering formation and potentially scaling up alternatives. For example, since education for private-sector production either for low-priced mass consumption or military purposes remains the dominant practice, "strengthening engineering service" by "multiplying the projects" of engineers requires examining both how this came to be the case in the United States and what might constitute other potential ends. Helping students better understand their commitments to broader social projects, he maintains, can help overcome the self-limitations of engineering curricula that Gary experienced so acutely as a student himself.

Gary later learned it is one thing to develop a local initiative built on one's own vision and energy and quite another to scale it up sufficiently to be integrated into curricula at other institutions and the working lives of engineers. To these ends, he first developed a system for mentoring graduate students to teach *Engineering Cultures* and then wrote and videotaped twenty-eight lectures from the course, nineteen of which are available for free as streamed, multimedia presentations. In 2005, Gary expanded the core pedagogical method in the course into a redefinition of engineering practice as "problem definition and solution." He had realized that mapping the locations, forms of knowledge, and desires of stakeholders in engineering work was not a supplement to engineering practice but an integral part of it, as collaborative problem definition. The privilege of this new knowledge brought with it new responsibilities. These include linking the curricular activities more closely to a parallel effort to build a new field of research called Engineering Studies.

WHAT IS GLOBAL ENGINEERING EDUCATION FOR?

In adding identities beyond the borders of home countries, contributors to this volume encountered discrepant moments and experiences of incongruence that led them to see themselves and their knowledge differently. They learned to recognize that what they had always taken for granted in both professional and personal practices could have been different, and, indeed, frequently became different in the process. The big payoff in accepting and engaging incongruence was in learning to analyze their own locations, forms of knowledge, and desires in relation to those of others. The privilege of such learning introduced, in turn, the very possibility of scaling up other practices in engineering pedagogy in addition to, alongside, or in place of those that have been dominant. They became committed, in particular, to practices of international and global engineering education.

For the nine engineers, the key encounters with incongruence came after they had already built identities as engineers. For Rick Vaz and Joe Mook, following work activities abroad produced encounters with other social and political points of view and multiple nontechnical dimensions of engineering work that motivated dramatic changes in their career paths. Rick began insisting that quality engineering work necessarily includes attention to its social, political, and cultural dimensions, and Joe began insisting that international education can be an engineering specialty. For Margie Pinnell, Jim Mihelcic, Linda Phillips, and Les Gerhardt, the new identities they gained beyond the United States came not through international travel but by experiencing changes in their work activities that made them rethink and redefine their identities as engineers in the world. Margie found herself a technical manager of work producing materials for making war. She redirected her travels to include supporting a program designed to help advance peace. Jim realized his focus on environmental remediation did not address the larger issue of the sustainability of life. He expanded the boundaries of environmental engineering to include it. Linda extrapolated years of construction work to an image of construction engineers paving green spaces everywhere. She redesigned education in senior design to expand what students consider professional practice, including both variable construction practices and new stakeholders. While working on the Apollo project, Les realized he had been trained to think of engineering as only a national resource. He began devoting himself to insuring students define engineering service in broader terms. Anu's case is unique because identities inside and outside her country of origin were part of her developing identity as an engineer in the first place. After struggles with humanitarian work in her home country, she refocused on expanding the boundaries of work in environmental engineering to include what communities want.

Dan Hirleman and Alan Parkinson might appear at first glance to be working wholly within conventional boundaries of engineering work. Both are helping students compete for jobs with companies that have built multinational networks, and they link this work to the goal of advancing American competitiveness. Yet they too are redefining engineering through their work. In Denmark, Dan encountered something entirely new, the experience of being an outsider. His search is for ways of understanding and doing engineering *with* other engineers and stakeholders who think differently, not *against* them. Alan established global competency as a strategic goal for his college in the interest of competitively attracting students, placing graduates, and empowering a country threatened by China. Actually supporting programs led him, however, to redefine his understanding of engineering service and engineering work to include humanitarian goals and activities.

The non-engineers and hybrids all found working to redefine the identities of engineers to also solve identity problems of their own. Bernd Widdig, Phil McKnight, and John Grandin all resisted the practice of separating literature from everyday life in the discipline of Foreign Language and Literature. Bernd accepted the risk of appearing to produce an "in-house Berlitz school" in order to connect his work with the realities of the present and build an internship program to send MIT students to Germany. Phil worked to build language programs for professionals at two institutions with the goal of producing knowledge workers who "will understand how to employ technology, science, and the humanities in ways that have a principled, positive impact on society and the world in which we live." John built a local initiative and then facilitated a national movement in international and global engineering education to insure that students gain "not only knowledge of the language" but understanding of "the perspectives of people from nations other than our own." Mike Nugent saw himself reproducing the separation between language learning and content learning that had so plagued him in earlier years, despite watching European professionals evidently achieve full integration of the two. For him, engineering education provided both an opportunity and a test site for funding programs that would achieve in the United States what he had seen in Europe. Gayle Elliott took the separation of work life and home life for granted as she worked her way up in administrative positions within a university organization. Realizing she could help students while making use of the knowledge she had gained through international travel solved a problem she did not know she had. She had come upon a way to connect together her personal and her work identities, producing a vocation in the process.

Finally, Juan Lucena and Gary Downey concentrated on using their non-engineer identities to inform the engineering identities of both themselves and others. Juan has worked to redefine engineering by making visible the power relations and politics in engineering and by helping "committed people trying to construct alternatives to engineering for military and free-market development." Gary has built a program of teaching and research to integrate into the heart of engineering education and training the practice of mapping the locations, forms of knowledge, and desires of stakeholders in engineering work.

For participants in this project, encountering conflicts between work identities and personal identities in experiences outside home countries frequently had the jarring effect of introducing new dimensions of choice. What they had previously taken for granted as their knowledge about themselves and their world became the product of their own experiences and judgments, their own decisions. An established identity and associated forms of knowledge became a point of view. And as at least some of what was given became variable, subject to chance and change, futures became more obviously dependent on questions and decisions in the present. What are my responsibilities? What do I want for myself? What do I want to do for others?

One set of concerns discussed in the workshop about practices of international and global engineering education workshop points out that little attention has been given to their effects on partners in other countries. "Almost all the articles [on international and global engineering education] I have reviewed to date," points out Anu Ramaswami in her contribution, "have addressed scale of the input in terms of numbers of interns or student participation. What about the scale of the output? What impacts are these programs having on the world outside of academia?" It is indeed essential to study and debate the effects international and global engineering education have not only on participants, future employers, and the abilities of home institutions to recruit good students but also on those hidden or made less visible by the dominant image or images to which one subscribes. It means inquiring into the local economy of home-stays for study-abroad students. It means asking what exchanges mean for host educational institutions or internships for host companies. It also means inquiring into the actual machinations of humanitarian, development, and sustainability projects, taking care to map flows of benefits and costs both inward and outward. It matters greatly where a program and its student participants end up. A significant body of research now exists, for example, critically examining the participation of engineers in international development.[35]

It also matters, at the same time, where students start and in what directions they head. Practices of education are in part about movement. They are about learners and those who lead them starting off somewhere and then heading off in particular directions. They are about acquiring identities, gaining knowledge and normative commitments along the way. For this reason, to understand and assess practices of international and global engineering education, it is crucial to include attention to where students start and the contents and directions of the steps they take. It is also crucial to make visible the types of questions they may be less likely to ask about themselves and their work if their education did not include adding identities outside the country.

Reporting on an ethnographic study, the education researcher Nadine Dolby found that the experiences of a group of U.S. students (non-engineers) studying abroad in Australia contrasted with what was promised by their study abroad office. The advertising for study abroad had "foreground[ed] the importance of the 'cross-cultural' experience," emphasizing the opportunity to learn about others. It was about "immersing oneself" in the foreign.[36]

What she found was that the "critical encounter of study abroad" for these affluent, white American students was actually with themselves. It was with their own identities. Finding themselves for the first time in a situation where "students become 'others,'" they found their American identity "invigorated" to an unprecedented and surprising extent, meaning it became visible to them. They found out that Australians had well-developed ideas about Americanness that did not necessarily fit their own understandings. Adding the identity of "'other' in Australia," they found themselves questioning the relation between the state and the country at home, exhibiting responses ranging from narrow defensiveness to a novel "cosmopolitan patriotism." As her title indicates, Dolby concluded

---

[35] Baillie, Engineers within a Local and Global Society, 2006; Catalano, Engineering Ethics, Peace, Justice, and the Earth, 2006; Downey and Lucena, "Engineers and Development," 2010b; Lucena et al., Engineering and Sustainable Community Development, 2010; Nieusma and Riley, "Designs on Development," 2010; Riley, "Resisting Neoliberalism in Global Development Engineering," 2007; Riley, Engineering and Social Justice, 2008; Zemach-Bersin, "American Students Abroad Can't Be 'Global Citizens'," 2008.

[36] Dolby, "Encountering an American Self," 2004, 150.

that study abroad for these students was in significant part an experience "encountering an American self."[37]

Coming to international and global engineering education from different directions, the contributors to this volume have all effectively become engineering educators with distinctive normative commitments and practices. Their trajectories are important and instructive examples of what might be called "encounters with an American engineering self." What might it take to embed such encounters systematically in the education and training of engineers?[38]

Toward the end of the workshop, John Grandin announced, "It's fine with me" if engineering students "are warmed" to international and global engineering education because they think it will "open up new career opportunities" and "help with American competitiveness and so forth." Engineering students tend to be "pragmatic," he said, meaning they tend not to ask questions about the boundaries of their identities and work as engineers.

Then leaning back in his seat and folding his arms with an air of confident expectation, he described what typically happens when students come back. "They'll come into my office and say, 'Dr. Grandin, I learned things that I never had any idea I was going to be learning'." They go on to explain in particular what they learned about themselves in relation to others they could now see and begin to understand. Invariably, he said, it is "a self eye-opening experience." Because "students can tell you at the end of the experience" what they can now see and how differently they understand themselves in the world, "I don't worry about it at the beginning."

Yet will the practices of reflecting critically on their knowledge and normative commitments ever become more than supplements to engineers' learning if their curricula do not "worry about" and address it at the beginning, middle, and end?

ACKNOWLEDGEMENTS

The research reported here was supported by NSF Grant #DUE-0752915. Thanks to members of the Virginia Tech Research in Engineering Studies group for comments on an earlier draft, including Jongmin Lee, Jacob Moore, Matt Wisnioski, and, especially, Kacey Beddoes.

REFERENCES

Abell, Peter. (2004) "Narrative Explanation: An Alternative to Variable-Centered Explanation?" *Annual Review of Sociology* 30: 287–310. DOI: 10.1146/annurev.soc.29.010202.100113 10

---

[37] Dolby, "Encountering an American Self, 2004, 162, 163, 172. Notably, such was not the case for Australians studying in the United States. In a follow-up study that describes the experiences of Australian undergraduates abroad in the United States, Dolby found that the Australians already had developed a "networked identity" for themselves. That is, they already located Australia (and themselves as Australians) in relation to other countries. They had already been plotting career trajectories that positioned the United States as a site for educational and job opportunities. Their major struggles lay not in finding out they held distinctively Australian points of view but in contrasting understandings of race in the United States and Australia. Dolby, "Globalisation, Identity, and Nation," 2005, 108.

[38] See Epilogue for one attempt.

Adams, Robin, Cheryl Allendoerfer, Tori Rhoulac Smith, David Socha, Dawn Williams, and Ken Yasuhara. (2007) "Storytelling in Engineering Education." Paper presented at the American Society for Engineering Education 2007 Annual Conference and Exposition. 11

American Society for Engineering Education. (1987) "A National Action Agenda for Engineering Education." Washington, D.C.: American Society for Engineering Education. 5

Atkinson, Paul. (1990) *The Ethnographic Imagination: Textual Constructions of Reality.* London: Routledge. DOI: 10.1525/aeq.1993.24.1.05x1739w 10

Atkinson, Paul, Amanda Coffey, Sara Delamont, and John Lofland. (2002) *Handbook of Ethnography.*London: Sage. 10

Baillie, Caroline. (2006) *Engineers within a Local and Global Society.* San Rafael, CA: Morgan & Claypool Publishers. DOI: 10.2200/S00059ED1V01Y200609ETS002 6, 39

Bell, D. (1991) "Insignificant Others: Lesbian and Gay Geographies." *Area* 23:323–329. 8

Bruner, Edward M. (1997) "Ethnography as Narrative." In *Memory, Identity, Community: The Idea of Narrative in the Human Sciences*, edited by Lewis P. Hinchman and Sandra Hinchman, 264–280. Albany, NY: SUNY Press. 10

Catalano, George. (2006) *Engineering Ethics, Peace, Justice, and the Earth.* San Rafael, CA: Morgan & Claypool Publishers. DOI: 10.2200/S00039ED1V01Y200606ETS001 6, 39

Committee on Prospering in the Global Economy of the 21st Century: An Agenda for American Science and Technology, Committee on Science, Engineering, and Public Policy, National Academy of Sciences, National Academy of Engineering, and Institute of Medicine of the National Academies. (2007) *Rising above the Gathering Storm: Energizing and Employing America for a Brighter Economic Future.* Washington, D.C.: The National Academies Press. 6

Cresswell, T. (1996) *In Place/out of Place: Geography, Ideology, and Transgression.* Minneapolis, MN: University of Minnesota Press. 8

Dolby, Nadine. (2004) "Encountering an American Self: Study Abroad and National Identity." *Comparative Education Review* 48, no. 2:150–173. DOI: 10.1086/382620 39, 40

Dolby, Nadine. (2005) "Globalisation, Identity, and Nation: Australian and American Undergraduates Abroad." *Australian Educational Researcher* 32, no. 1:101–118. 40

Downey, Gary Lee. (1992) "CAD/CAM Saves the Nation?: Toward an Anthropology of Technology." *Knowledge and Society* 9:143–168. 5

Downey, Gary Lee. (1995) "The World of Industry-University-Government: Reimagining R&D as America." In *Technoscientific Imaginaries*, edited by George Marcus, 197–226. Chicago, IL: The University of Chicago Press. 5

Downey, Gary Lee. (1998) "Introduction." In *The Machine in Me: An Anthropologist Sits among Computer Engineers.* New York, NY: Routledge. 5, 12

Downey, Gary Lee. (1998) *The Machine in Me: An Anthropologist Sits among Computer Engineers.* New York, NY: Routledge.

Downey, Gary Lee. (2007) "Low Cost, Mass Use: American Engineers and the Metrics of Progress." *History and Technology* 22, no. 3:289–308. DOI: 10.1080/07341510701300387 5

Downey, Gary Lee. (2009) "What Is Engineering Studies For?: Dominant Practices and Scalable Scholarship." *Engineering Studies: Journal of the International Network for Engineering Studies* 1, no. 1:55–76. DOI: 10.1080/19378620902786499 13

Downey, Gary Lee. (2010) "Location, Knowledge, and Desire: From My Two Conservatisms to *Engineering Cultures* and Countries." In *What Is Global Engineering Education For?: The Making of International Educators*, edited by Gary Lee Downey and Kacey Beddoes. San Rafael, CA: Morgan & Claypool Publishers. 8

Downey, Gary Lee and Juan C. Lucena. (2010) "Engineers and Development: The Need for Engineering Studies." In *Engineering: Issues and Challenges for Development*, edited by Tony Marjoram. Paris, France: UNESCO Publishing. 39

Franzosi, Roberto. (1998) "Narrative Analysis—or Why (and How) Sociologists Should Be Interested in Narrative." *Annual Review of Sociology* 24:517–554. DOI: 10.1146/annurev.soc.24.1.517 10

Friedman, Thomas L. (2005) *The World Is Flat: A Brief History of the Twenty-First Century.* New York, NY: Farrar, Straus and Giroux. 6

Goldman, Steven L., (1989) *Science, Technology, and Social Progress.* Bethlehem, PA: Lehigh University Press. 5

Latour, Bruno. (1987) *Science in Action: How to Follow Scientists and Engineers through Society.* Cambridge, MA: Harvard University Press. 7

Law, John. (2004) *After Method: Mess in Social Science Research.* London and New York: Routledge. 9

Lengwiler, Martin. (2008) "Participatory Approaches in Science and Technology: Historical Origins and Current Practices in Critical Perspective." *Science, Technology, & Human Values* 33, no. 2:186–200. DOI: 10.1177/0162243907311262 8

Lucena, Juan. (2005) *Defending the Nation: U.S. Policymaking to Create Scientists and Engineers from Sputnik to the 'War against Terrorism'.* Lanham, MD: University Press of America, Inc. DOI: 10.1007/s11024-007-9062-y 5

Lucena, Juan, Jen Schneider, and Jon A. Leydens. (2010) *Engineering and Sustainable Community Development.* San Rafael, CA: Morgan & Claypool Publishers. DOI: 10.2200/S00247ED1V01Y201001ETS011 39

McLoughlin, Lisa. (2009) "Success, Recruitment, and Retention of Academically Elite Women Students without STEM Backgrounds in U.S. Undergraduate Engineering Education." *Engineering Studies* 1, no. 2:151–168. DOI: 10.1080/19378620902911592 17

MIT Commission on Industrial Productivity. (1989) "Education and Training for the United States: Development the Human Resources We Need for Technological Advance and Competitiveness." Cambridge, MA: MIT Press. 5

National Academy of Engineering. (2005) *Educating the Engineer of 2020: Adapting Engineering Education to the New Century.* Washington, D.C.: The National Academies Press. 6

National Science Board. (1988) "The Role of the National Science Foundation in Economic Competitiveness." Washington, D.C.: National Science Board. 5

Nieusma, Dean and Donna Riley. (2010) "Designs on Development: Engineering, Globalization, and Social Justice." *Engineering Studies* 2, no. 1:29–59. DOI: 10.1080/19378621003604748 39

Nowotny, H., P. Scott, and M. Gibbons. (2001) *Re-Thinking Science: Knowledge and the Public in an Age of Uncertainty.* Cambridge, UK: Polity. 8

Okely, Judith. (1992) "Anthropology and Autobiography: Participatory Experience and Embodied Knowledge." In *Anthropology and Autobiography*, edited by Judith Okely and Helen Calllaway, 1–29. London and New York: Routledge (Taylor & Francis Group). 11

Patterson, Molly and Kristen Renwick Monroe. (1998) "Narrative in Political Science." *Annual Review of Political Science* 1:315–331. DOI: 10.1146/annurev.polisci.1.1.315 10

Riley, Donna. (2007) "Resisting Neoliberalism in Global Development Engineering." Paper presented at the 2007 American Society for Engineering Education Conference & Exposition. 6, 39

Riley, Donna. (2008) *Engineering and Social Justice.* San Rafael, CA: Morgan & Claypool Publishers. DOI: 10.2200/S00117ED1V01Y200805ETS007 6, 39

Silverstone, R., E. Hirsch, and D. Morley. (1992) "Information and Communication Technologies and the Moral Economy of the Household." In *Consuming Technologies: Media and Information in Domestic Spaces*, edited by R. Silverstone and E. Hirsch, 15–31. London: Routledge. DOI: 10.4324/9780203401491_chapter_1 8

Sommerville, P. (1992) "Homelessness and the Meaning of Home: Rooflessness or Rootedness?" *International Journal of Urban and Regional Research* 16:528–539. 8

The National Commission on Excellence in Education. (1983) "A Nation at Risk: The Imperative for Educational Reform. A Report to the Nation and the Secretary of Education, U.S. Department of Education." Washington, D.C.: The National Commission on Excellence in Education. 5

Valentine, Gill. (1998) "'Sticks and Stones May Break My Bones': A Personal Geography of Harassment." *Antipode* 30, no. 4:305–332. DOI: 10.1111/1467-8330.00082 8

Wasburn, Mara H., Adrienne R. Minerick, and Valerie L. Young. (2009) "Mothers on the Tenure Track: What Engineering and Technology Faculty Still Confront." *Engineering Studies* 1, no. 3:217–235 DOI: 10.1080/19378620903183530 17

Wisnioski, Matthew. (2009) "Liberal Education Has Failed: Reading Like an Engineer in 1960s America." *Technology and Culture* 50, no. 4:753–782. DOI: 10.1353/tech.0.0346 12

Zemach-Bersin, Talya. (2008) "American Students Abroad Can't Be 'Global Citizens'." *Chronicle of Higher Education* 54, no. 26:A34. 39

Zuiderant-Jerak, Teun and Casper Bruin Jensen. (2007) "Editorial Introduction: Unpacking 'Intervention' in Science and Technology Studies." *Science as Culture* 16, no. 3:227–235. DOI: 10.1080/09505430701568552 8

CHAPTER 2

From Diplomacy and Development to Competitiveness and Globalization: Historical Perspectives on the Internationalization of Engineering Education

Brent K. Jesiek and Kacey Beddoes

INTRODUCTION

This chapter examines the historical development of global engineering education in the United States from roughly the 1940s to the present, with emphasis on programs and initiatives designed to provide American engineering students with opportunities to study or work abroad. The paper is organized around overlapping historical periods, each with partially distinct justifications for expanding these kinds of opportunities. The first section traces trends from the World Wars into the Cold War, with particular focus on the 1940s to 1970s. In addition to examining the historical origins and expansion of study abroad in the U.S., this part focuses on "diplomacy" and "development" as key drivers for the early internationalization of engineering education. In a second major section, we examine how arguments about enhancing America's economic competitiveness emerged as an important new justification for internationalizing engineering education, especially during the 1980s. Finally, in a concluding section, we review some contemporary developments, including the rise of globalization discourses as a powerful new driver and rationale for expanding global engineering education in the United States.

Throughout our analysis, we also highlight a range of other motivations for global engineering education that have surfaced over the decades, including benefits to participants. We also point to some factors common to programmatic success, such as committed individual students and dedicated faculty members. And throughout our account, we emphasize that many of the persistent challenges and obstacles associated with scaling up global engineering education have longer histories dating back some five decades.

This chapter is mainly historical in character, and it is significantly based on extensive archival research. Our in-depth and systematic examination of publications of the American Society for Engineering Education (ASEE) – including *Journal of Engineering Education* (or *Engineering Education*, as it was known for some time), *TECHNOS*, and *ASEE Annual Conference Proceedings* – was supplemented by targeted keyword searches in databases like *Engineering Village* and *Google Scholar*. As specific topics of interest were identified, we also searched for and incorporated additional source material, including original journal articles and conference papers, special reports, secondary historical accounts and summaries, and publicly available biographical information. While our goal is to present a reasonably comprehensive overview of major historical trends in global engineering education, this account is necessarily patchwork in character at the level of individual actors and programs.

We hope our analysis inspires, and provides foundations for, further research.

PART 1 – INTERNATIONALIZING ENGINEERING EDUCATION FROM WORLD WARS TO COLD WAR

A NEW KIND OF CULTURAL DIPLOMACY: INTERNATIONAL EDUCATION AND EXCHANGE

For many centuries, students and professors have travelled far and wide to visit, study, research, and teach. But for American undergraduate students, education abroad has more recent origins. In fact, no less a figure than Thomas Jefferson argued against sending American students to Europe, warning that it could lead a young man to lose touch with his native knowledge, morals, habits, and even happiness.[1] In the Nineteenth century, a new wave of commentators – including some prominent university presidents – made similar arguments against international education, while adding that American institutions were better and cheaper than schools abroad.[2]

Yet for a number of reasons, international education gained new appeal in the wake of World War I. To begin, some Americans who spent considerable time in Europe during and after the war became advocates for study abroad, which they saw as providing the kinds of language and cultural training needed for enhanced career success and upward mobility, especially for foreign language majors.[3] Still other period commentators, like University of Delaware President Walter Hullihen, recognized emerging business and government needs. Advocating the school's foreign

[1] Schwaneger, "The Junior Year Abroad," 1970, 155.
[2] Schwaneger, "The Junior Year Abroad," 1970, 155.
[3] Walton, "Internationalism and the Junior Year Abroad," 2005, 259–263.

study programs – which expanded impressively in the 1920s and 1930s, especially in France – Hullihen explained: "[O]ne of our specific aims is to create…a great reservoir of college trained business men upon whom commerce and government may draw for work that involves a knowledge of the language and customs of other countries."[4]

Other initiatives, such as the 1919 founding of the Institute for International Education (IIE), were significantly focused on international goodwill and understanding.[5] In addition to being supported by the Carnegie Endowment for National Peace, IIE inaugural director Stephen P. Duggan declared in 1920: "The institute believes that it is as essential for Americans to know of the difficulties and problems of other countries as for the people of other countries to know something about us, in order that international good-will may be realized."[6] The IIE undertook a wide variety of activities in subsequent years, including as a leading nongovernmental advocate and administrator for the exchange of university students and faculty between the U.S. and other countries, and especially those in Europe.[7]

In the 1930s and 1940s, U.S. policymakers similarly started seeing educational exchange as a valuable form of "cultural diplomacy." Beginning in 1937, for instance, the U.S. government began subsidizing student and faculty exchanges with Latin America as part of Roosevelt's Good Neighbor Policy.[8] Other programs from around this time, such as the Fulbright Act of 1946 and U.S. Information and Education Exchange Act (or Smith-Mundt Act) of 1948, had similar diplomatic intentions and supported student exchanges in an even wider range of academic fields and destination countries.[9]

As funding for study abroad increased, skepticism about the value of such experiences started to wane and a dramatic expansion of international education followed. The number of study abroad programs at U.S. universities increased from six in 1950 to more than a hundred by the late 1950s.[10] Further, IIE reports show that the number of American students abroad increased from about 9,500 in 1954-1955 (the first year these data were collected) to more than 15,000 in 1959-1960.[11] Many of the study abroad programs active during this period were specifically focused on the humanities or medical fields, while others were open to all majors.

ENGINEERING STATESMANSHIP AND BUSINESS ABROAD: EMERGENT NEEDS AND DESIRABLE TRAITS

Beginning in the 1930s, growing numbers of American engineers and other technical experts were going abroad to support foreign aid initiatives. As one early example of this trend, the U.S. devel-

[4] Walton, "Internationalism and the Junior Year Abroad," 2005, 261.

[5] Walton, "Internationalism and the Junior Year Abroad," 2005, 261.

[6] Walton, "Internationalism and the Junior Year Abroad," 2005, 261.

[7] Ferguson, "75 Years of the Junior Year in Munich," 2007, 124; IIE, "A Brief History of IIE," n.d., 2010; Schwaneger, "The Junior Year Abroad," 1970, 155–156.

[8] Miller, "'An Effective Instrument of Peace'," 2006; Walton, "Internationalism and the Junior Year Abroad," 2005, 278.

[9] Bu, "Educational Exchange and Cultural Diplomacy in the Cold War," 1999, 399; Johnson, *1600 Pennsylvania Avenue*, 1960, 220–221; Miller, "'An Effective Instrument of Peace'," 2006.

[10] Schwaneger, "The Junior Year Abroad," 1970, 156; Walton, "Internationalism and the Junior Year Abroad," 2005, 278.

[11] IIE, *Open Doors 1955-56*, 1956, 13; IIE, *Open Doors 1961*, 1961, 17.

oped bilateral technical assistance programs to increase the production of foodstuffs and materials during World War II, especially in Latin America.[12] The U.S. also gave technical assistance to many European countries after the Second World War, especially via the Marshall Plan.[13]

Early Marshall Plan successes informed the launch of Truman's ambitious "Point Four" initiative in 1949. Per Truman, this was a "bold new program for making the benefits of our scientific advances and industrial progress available for the improvement and growth of underdeveloped areas."[14] He added that sharing American technical know-how could nurture peace, democracy, and freedom via economic development. By 1951, there were already "more than 400 technicians working in some 30 countries in Latin America, Africa, the Middle East and Asia on Point 4 projects."[15] And in 1955, 635 American technicians were in Latin America alone, with 200 vacancies remaining.[16] Most served as advisors, consultants, and educators, working on projects in areas such as agricultural production, irrigation, food processing and storage, water and sewer systems, geological survey, mining development, and transportation infrastructure.[17] Inter-university contracts between U.S. institutions and their counterparts abroad emerged during this period as one mechanism for providing technical assistance.[18]

Many American companies were also starting to employ notable numbers of engineers abroad in the immediate post-war period, including to work on development projects and support branch operations. As one 1949 article noted, "A large segment of U.S. business is becoming more internationally minded," in part due to growing "national interdependencies."[19]; Additionally, an article published the same year showed that ten US firms employed 350 engineers abroad, with more than 200 of these in Latin America alone.[20] The author added that many of these companies felt their employees should gain extensive domestic experience before going abroad. Employers identified recruitment as a particular challenge, and five indicated a preference for hiring foreign nationals educated in the U.S. rather than American born graduates.[21]

In the early 1950s, a series of commentators discussed how to best prepare American engineers for such assignments abroad. In 1951, for instance, Pacific Power and Light President Paul B. McKee recommended that a young engineer should "serve his apprenticeship at home" prior to foreign service, which would include a high-quality yet conventional university education and substantial domestic work experience.[22] But McKee also acknowledged that this kind of work demanded a "human outlook," "lack of prejudice," "desire to understand and get along with people," and knowledge

---

[12]NPA, "Technical Cooperation," 1955a, 1.

[13]Willard, "Europe Looks to America for Technological Assistance," 1948.

[14]Truman, "Truman Inaugural Address," 1949.

[15]Bennett, "The Point 4 Program," 1951, 139.

[16]NPA, "The Role of Universities in Technical Cooperation," 1955b, 17.

[17]Allen, "Where Do We Stand on Point Four?" 1950; Bennett, "The Point 4 Program," 1951; Long, "Designing Engineering Education for Development Needs," 1975, 66; NRC, "The Role of U.S. Engineering Schools in Development Assistance," 1976.

[18]Bjork, "Engineering Education in an International Perspective," 1960.

[19]Hughlett, "American Engineering and Industry Abroad," 1949, 571, 575.

[20]Thompson, "International Contacts for Engineering Educators," 1949, 138.

[21]Thompson, "International Contacts for Engineering Educators," 1949, 138–139.

[22]McKee, "The Role of the Engineer in International Affairs," 1951, 155.

of local language, laws, and customs.[23] With proper preparation, McKee added, American engineers could serve as "ambassadors of good will," thereby helping "win the game against the forces which are trying to take our position of leadership from us."[24] While McKee was not specific about this threat, his remarks reflected intensified anti-Communist sentiment in the U.S. during the late 1940s and into the 1950s.[25]

McKee also had little to say about changing engineering education to cultivate future engineer-ambassadors, but others had more progressive visions. S. S. Steinberg, Dean of Engineering at the University of Maryland, worked in these directions in 1949 as he reflected on opportunities and challenges associated with providing technical assistance to developing countries, especially via the Point Four program. "[I]t may be that supplying this technical knowledge" Steinberg explained, "will require training in engineering colleges fully as broad in the technical fields as now, but probably with a somewhat broader basis in the humanistic-social field, aiming toward living and working in foreign areas."[26] Also discussing the Point Four program in an address delivered the same year, Assistant Secretary of State George Allen added: "[S]ome of our engineering students, to be properly prepared for their work, should have foreign languages and some 'area studies'."[27]

Similar comments about the need to broadly train engineers as citizens and statesmen continued to surface in the 1950s as more American engineers went abroad and challenges associated with recruiting and preparing this new kind of expert became more apparent. In 1952, for example, Cornell Professor and head of Civil Engineering N. A. Christensen lamented that few civil engineering graduates "have a broad enough foundation to support a career in important international relations requiring extensive knowledge and training," and he added that few schools could train the so-called "engineering statesmen."[28] Around the same time, consulting engineer Morris Cooke concluded that "the college world is pretty well convinced that engineers must take more of a hand in the world of affairs lying entirely outside of our technical domain. Engineering *students* are our best prospects."[29] He also alluded to difficulties related to maintaining traditions in engineering education while introducing new pedagogical approaches.

RISKY UNDERTAKINGS: AMERICAN ENGINEERING STUDENTS GO ABROAD IN THE 1950s

As the preceding section suggests, growing consensus about the need to prepare more American engineers for work abroad was accompanied by a lack of agreement about the best pathways for training such professionals. At the same time, relevant educational experiences like study abroad were marginal or invisible in most engineering schools. These trends are striking for a number of reasons. As noted above, study abroad expanded considerably during the 1950s. Additionally,

[23] McKee, "The Role of the Engineer in International Affairs," 1951, 157–158.
[24] McKee, "The Role of the Engineer in International Affairs," 1951, 160.
[25] Zinn, "A People's History of the United States," 1980, Ch. 16.
[26] Steinberg, "Engineering Know-How for the World," 1948, 411.
[27] Allen, "Where Do We Stand on Point Four?" 1950, 69.
[28] Christensen, "Social and Political Forces Affecting Civil Engineering Education," 1952, 429, 431.
[29] Cooke, "The Role of the Engineer in Community Affairs," 1951, 74.

growing numbers of engineering faculty and administrators were supporting an expanding array of international activities at their own institutions, including inter-university contracts with partner institutions abroad.[30]

Nonetheless, these trends were not indicative of a more general internationalization of engineering education. According to data reported by IIE, engineering majors represented just over 1% of all American students studying abroad for credit through the mid and late 1950s.[31] Canada was their top destination, followed by Western Europe and Latin America.[32] Many factors probably contributed to this low level of participation, including a long-standing perception that international experiences were more appropriate for students in other majors. Domestic demand for engineers was also strong in the 1950s, making it less likely that graduates would seek out employment abroad.[33] Others invoked the old argument that it made little sense for students to turn away from excellent educational opportunities available in the States. One early advocate for international education summarized the prevailing view among engineering deans: "Why should our men desire to enroll in graduate engineering courses abroad when it is most unusual to find laboratory and library facilities or staffs comparable to those in the U.S.?"[34]

While few international programs specifically targeted engineering students in the 1950s, two exceptions are worth noting. The Antioch Education Abroad (AEA) initiative was launched in 1956, and in 1958, it initiated a new exchange program in partnership with the Carl Duisberg Foundation of Germany and U.S.-based Ford Foundation.[35] It allowed Antioch business and engineering students to spend a year in Germany to take courses and work in industry. According to one account, Antioch's innovative programs were at the time viewed as a "risky undertaking," especially because "including work experience in the educational program abroad was practically unheard of."[36] While about 1,500 students had participated in AEA program offerings by the late 1960s, information about the experiences of the engineering students is scarce.[37]

A second notable example is the International Association for the Exchange of Students for Technical Experience (IAESTE), which was founded in Western Europe in 1948 to arrange cross-national exchanges of students into technical positions, mainly in the private sector.[38] By around 1960, it claimed twenty-six member countries, with 3,000 companies providing training opportunities for more than 6,000 students.[39] While initially centered in Europe, IAESTE's U.S. operations expanded in the 1950s, and in 1959, the organization successfully placed ninety-six

---

[30] Andrus, "Technical Assistance through Inter-University Contracts," 1956; Bibber, "The Background of 'Sister University' Contracts Between American and Foreign Universities," 1956; Bjork, "Engineering Education in an International Perspective," 1960.
[31] IIE, *Open Doors 1955-56*, 1956, 43; IIE, *Open Doors 1957*, 1957, 43; IIE, *Open Doors 1961*, 1961, 34.
[32] IIE, *Open Doors 1955-56*, 1956, 43; IIE, *Open Doors 1957*, 1957, 43; IIE, *Open Doors*.
[33] Armsby, "An Estimate of Graduate Engineers and Engineering Jobs," 1958; Merrill, "The Scientific Manpower Shortage and Its Implications for the Engineering Technician," 1958.
[34] Thompson, "International Contacts for Engineering Educators," 1949, 137.
[35] Bjork, "Engineering Education in an International Perspective," 1960, 689; Institute of Research on Overseas Programs, *The International Programs of American Universities*, 1958, 83.
[36] Abrams, "The Impact of Antioch Education Through Experience Abroad," 1979, 176.
[37] Abrams, "The Impact of Antioch Education Through Experience Abroad," 1979, 178.
[38] Justice and Wischeidt, "Pan-American Exchanges in Engineering, Science, and Agriculture," 1961, 453.
[39] Justice and Wischeidt, "Pan-American Exchanges in Engineering, Science, and Agriculture," 1961, 453.

American students abroad, including seventy-three engineers.[40] Virtually all of these placements were in Europe. The organization also reported strong interest from a larger number of students, but they noted exchanges were limited by both the lack of a "working knowledge of a foreign language" among U.S. engineering students and a shortage of reciprocal openings in U.S. companies.[41]

Despite low levels of participation and a dearth of programs targeted to engineering students, other evidence suggests changing attitudes toward study abroad among some within the engineering community. In 1960, for example, a distinguished panel at an IIE conference concluded that: "more Americans should be studying engineering abroad,"[42] and added that undergraduate engineering students could especially benefit from: (1) a strong emphasis on mathematics by European technical schools; (2) opportunities to compare how Europeans and Americans handle the same problems; and (3) "the broadening experience of study abroad on the engineer as a world citizen."[43] They also reiterated common obstacles, including a lack of language expertise in the engineering student population and a shortage of industry sponsorship for exchange programs.

Nonetheless, some forward-looking engineering schools did manage to expand their international profile in the 1950s, 60s, and 70s, providing American students and faculty with new opportunities for experiences abroad. These ranged from traditional study abroad programs for undergraduate students to international research initiatives for graduate students and faculty. In the sections that follow, we review several notable programs and initiatives.

NEW OPPORTUNITIES FOR STUDY AND RESEARCH ABROAD IN THE 1960s AND 1970s

"Slide Rule Amigos" ran the headline of a 1963 article about a novel exchange program between the University of Wisconsin-Madison (UWM) and Mexico's Instituto Tecnológico y de Estudios Superiores de Monterrey (Monterrey Tec).[44] After noting the presence of many foreign engineering students in the U.S., the article explained: "It is highly unusual for an American undergraduate to travel to a foreign university in order to pursue an engineering career." Echoing earlier commentators, the author added that the high quality of American engineering schools, coupled with a lack of foreign language training, meant little interest in study abroad among engineering students. The "Wisconsin-Monterrey Program" profiled in the article was therefore exceptional indeed, given that it aimed to send American engineering students to Mexico.

The origins of the initiative can be traced back to at least 1959, when Monterrey Tec's President approached his counterpart at UWM about launching an exchange program.[45] While the idea was supported by the UWM administration, the attractiveness of the proposal was also enhanced by other factors, including various pre-existing relationships between the two schools, the College's prior track

[40] Crossley, "Foreign Summer Jobs," 1958; "Do You Know...?" 1961, 260.
[41] "Do You Know...?" 1961, 260.
[42] Shank, "Exchange Education in Engineering," 1960, 329.
[43] Shank, "Exchange Education in Engineering," 1960, 329.
[44] Raducha, "I'll Remember This Trip," 2008, 27.
[45] Raducha, "I'll Remember This Trip," 2008, 19; Wengler, "UW's International Engineering Champion," 1981, 7.

record with inter-university contracts, and a modest grant from the Carnegie corporation to help determine "whether an exchange program for engineers could work in addition to the programs for liberal arts students."[46]

Beginning in 1961-1962, American students could spend their junior year at Tec for language training, regular academic work, and supplemental travel. In its early years, Wisconsin and Case Institute (now Case Western Reserve University) partnered to send seven or eight students from each school to Mexico, and the program quickly stimulated other student and faculty exchanges between Wisconsin and Monterrey Tec.[47] The University of Wisconsin-Milwaukee, Carnegie Mellon, Stanford, and Cornell also sent students to Monterrey Tec in the 1960s as part of the same initiative, and by 1965, the program claimed forty-two prior American participants.[48]

Yet the viability of the initiative was in question by the mid-1960s, especially as Carnegie funding ran out and American student applicants remained sparse.[49] As a 1965 description of the program explained: "[E]ngineering students are not traditionally oriented toward foreign study, and ... are usually engrossed in meeting their technical course requirements."[50] However, the program rebounded to become the College's flagship study abroad initiative for more than twenty years, and by the late 1970s, more than 100 UWM engineering students had studied at Tec.[51]

The success and longevity of the program can be attributed to many factors. It is notable, for example, that Tec was accredited by a U.S. agency and widely recognized as providing quality technical education, so concerns about the rigor of the school's coursework and difficulties related to transferring credits were minimized.[52] Demand for former participants was also strong, especially from American firms with operations in Mexico.[53] A sustained mix of internal and external funding, strong leadership within the College, and solid administrative support were also important ingredients in the program's success.

Other innovative programs from the 1960s and 1970s provided students with opportunities to participate in various kinds of international activities, including cross-national research projects. One early initiative of this type was MIT's Inter-American Program in Civil Engineering. It was founded in 1961 by Charlie Miller, head of Civil Engineering, who was inspired by his own work experiences as a consultant in Puerto Rico.[54] With support from the Carnegie Foundation, Ford Foundation, and several federal agencies, the program engaged students in "real engineering activity by working with Latin American students in their own countries."[55] Stated program benefits included "significant

[46] Raducha, *"I'll Remember This Trip,"* 2008, 19.

[47] Education and World Affairs, *The University Looks Abroad*, 1965, 142–143.

[48] Crick, "International Opportunities for Student Engineers," 1970, 13; Raducha, *"I'll Remember This Trip,"* 2008, 19; Spychalla, "The Wisconsin-Monterrey Exchange Program," 1965, 14.

[49] Education and World Affairs, *The University Looks Abroad*, 1965, 142–143.

[50] Education and World Affairs, *The University Looks Abroad*, 1965, 143.

[51] Crick, "International Opportunities for Student Engineers," 1975, 1; Kusel, "A Change of Place," 1979; Raducha, *"I'll Remember This Trip,"* 2008, 46.

[52] Raducha, *"I'll Remember This Trip,"* 2008, 19; Crick, "International Opportunities," 1970, 13.

[53] Barry, "Foreign Schools Attract Students," 1973; Education and World Affairs, *The University Looks Abroad*, 1965, 143.

[54] McGarry, "The M.I.T. Inter-American Program in Civil Engineering," 1966; Miller and McGarry, "Inter-American Program," 1966, 11.

[55] McGarry, "The M.I.T. Inter-American Program in Civil Engineering," 1966.

and valuable" research activities, personal and professional growth for participants, and institution building.[56]

By 1966, a program report described thirteen associated research projects in Colombia, Mexico, Venezuela, Argentina, Brazil, Chile, and Peru.[57] Some of these were clearly linked to development concerns, including research on water resource management and testing structural responses to earthquakes. Other projects were not closely bound to local conditions or needs, except for the involvement of relevant experts and institutions abroad. Many American students conducted research through the program during their senior year, often as part of multi-national design teams.[58] However, the majority of participants (60% in 1972) came from Latin America to MIT.[59] The program declined and then ceased activity in the 1970s, but one of its lasting effects was to stimulate continued international activity among former participants, as we note below.

Also in the early 1960s, the Rural-Industry Technical-Assistance (RITA) program was launched at UCLA.[60] It was largely the brainchild of Morris Asimow, a passionate and visionary engineering professor with a penchant for undertaking atypical projects. Beginning in 1962, teams of UCLA professors and graduate students spent summers in the impoverished Brazilian state of Ceará performing feasibility studies and developing preliminary designs for small industries.[61] They worked jointly with faculty and students from the University of Ceará, while undergraduate students from UCLA got involved with the projects in their senior design courses. By 1965, the program had involved 150 seniors and thirty graduate students.[62] This group included many engineers, but student participants came from fields as diverse as business administration, economics, anthropology, and agronomy. Many Brazilian participants also visited UCLA for further education. By 1972, a total of seventy-seven U.S. graduate students had participated in the program, and eighteen of these had done research counting toward their degrees.[63]

RITA's stated goals and benefits included: (1) advancing the economic development of underdeveloped communities; (2) initiating activities that could be sustained by local stakeholders; (3) ensuring that U.S. funds supported programs that strengthened individuals and communities; and (4) generating new knowledge and enhancing student learning.[64] The specific educational benefits associated with RITA included improving the ability of student participants to: grapple with open-ended, real-world problems; work in groups; cross disciplinary boundaries; understand a foreign culture; and "accept that our approach to solving problems here in the United States will not

---

[56] McGarry, "The M.I.T. Inter-American Program in Civil Engineering," 1966.

[57] Miller and McGarry, "The Inter-American Program," 1966, 8–10.

[58] Smith, "Overseas Branches of American Universities," 1968, 271.

[59] Burr, "M.I.T. Involved in Three Major International Programs," 1972, 68.

[60] Asimow and McNown, "Engineering Education and International Development," 1965; Smith, "Overseas Branches of American Universities," 1968, 271–272. For further information on Asimow see: "Backland's Capitalism," 1965; and Lyman et al., "Morris Asimow," 1985, 13–14.

[61] "Backland's Capitalism," 1964; "UCLA Engineering," 2009 n.d.

[62] Asimow and McKnown, "Engineering Education and International Development," 1965, 67.

[63] Woodson and Gilkeson, "Technical Assistance in Communities in Northeast Brazil," 1972, 30.

[64] Asimow and McNown, "Engineering Education and International Development, 1965, 67–68.

necessarily solve similar problems in other parts of the world."[65] Program organizers also noted that the program could potentially improve faculty attitudes toward foreign students, connect foreign students with urgent needs back in their home countries, and positively influence the development of universities abroad.

However, early RITA successes in Ceará proved difficult to sustain and replicate as the program grew. As two observers summarized: "[S]erious conceptual and operational problems diminished anticipated accomplishments."[66] These same commentators pointed to a lack of long-term program evaluation to track successes and failures. And when an external evaluation did finally occur, one outcome was the elimination of U.S. students from teams in Brazil, "except in the cases of proven language and professional competence."[67] As one participant noted, "all of the United States team could be considered completely inoperative in Portuguese," and they had difficulties applying technology in industry.[68] As these statements suggest, enhancing the ability of U.S. participants to work abroad was largely overshadowed by RITA's focus on achieving development-related goals in Brazil, and even these were only partially successful.

TRAINING ENGINEERS FOR INTERNATIONAL SERVICE AND WORLD DEVELOPMENT

Programs like RITA once again suggested that preparing American engineers for practice abroad – and especially for work in developing countries – was a significant challenge. One early effort to train engineers for "international service" was launched at Michigan State University in 1960. Students in this five-year program received bachelor of science and bachelor of arts degrees. In addition to technical coursework in a specific area of engineering (chemical, electrical, mechanical, etc.), students took liberal arts courses "appropriate to the region of the world where they plan to work."[69] As one news item noted, the initiative "may go a long way toward eliminating the 'Ugly American' criticisms of overseas projects."[70] While little is known about the fate of this program, other initiatives with similar goals eventually followed. In the late 1960s, for example, the Purdue Fellows in Latin America (PFLA) program placed early career engineers in eighteen-month assignments with host employers in Argentina, Brazil, and Chile, with the goal of creating "an adequate corps of U.S. professionals capable of performing effectively in the economic, social and cultural environment of Latin America."[71]

While evidence suggests that the longevity and impacts of the Michigan State and Purdue programs were limited, discussions about desirable characteristics for development-oriented engineers continued to surface in the 1970s. In 1972, for instance, a World Bank representative noted that engineers working on development projects abroad could benefit from: (1) background knowledge

---

[65] Asimow and McNown, "Engineering Education and International Development, 1965, 68–69.
[66] Woodson and Gilkeson, "Technical Assistance in Communities in Northeast Brazil," 1972, 28.
[67] Woodson and Gilkeson, "Technical Assistance in Communities in Northeast Brazil," 1972, 34.
[68] Shen, "Technical Assistance in Northeast Brazil," 1974, 70–71.
[69] "Engineering for International Service," 1960.
[70] "Engineering for International Service," 1960.
[71] Lackner et al., "A Latin American Engineering Internship Program," 1970.

about social and economic context, (2) language training, and (3) an enhanced ability to communicate effectively in foreign environments.[72] Just a few years later, a Stanford affiliate presented a similar list of traits, including being able to: adapt Western methods and technologies to match local needs; understand a host country's history, government, culture, language, and customs; and avoid "loss of face" when working in different cultural contexts.[73]

By 1976, Joel Bernstein of the U.S. Agency for International Development (AID) estimated that 7,000 American engineers were working on projects abroad, including many as employees of U.S. firms. In his view, this meant "an increasing need for U.S. engineering schools to turn out graduates who understand something about the conditions governing industry in the less developed countries, as well as elsewhere overseas, and be able to analyze and advise on problems that arise in these countries."[74] Around this same time, another commentator aptly summarized that preparing engineers for development work "requires more than does conventional preparation."[75]

INTERNATIONAL ENGINEERING EDUCATION IN THE 1960s AND 1970s: FAILING TO SCALE

Some of the earliest U.S. programs offering experiences abroad for American engineers were linked to many different motivations and goals. As noted above, some initiatives were mainly concerned with preparing engineers for practice in developing contexts. The Wisconsin-Monterrey program was more generally framed as providing students with expanded career opportunities. IAESTE was also pitched as improving employability of participants, while at the same time promoting "international understanding and goodwill amongst the students of all nations."[76] The objectives of Antioch Education Abroad, on the other hand, included building each student's academic competence, "personal philosophy," and ability to intelligently participate in society – all reflecting the school's liberal arts tradition.[77]

During the 1960s and 1970s, a handful of other engineering schools, including Stanford and the University of Illinois at Urbana-Champagne (UIUC), were developing their own international offerings for engineering students, often motivated by a similarly diverse range of objectives.[78] Yet despite these pockets of activity, the number of engineering students abroad did not rise appreciably during this period. IIE data, for instance, show that the percentage of engineering students among all study abroad participants remained in the 1-2% range through the 1960s.[79] These figures peaked in 1968-1969, when it was estimated that 464 engineering students were among the more than 25,000

[72] Engelmann, "Engineers in Developing Countries," 1972, 66.
[73] Ireson, "Some Problems in Technical Assistance to LDC Universities," 1975, 21–22.
[74] Bernstein, "The U.S. Role in Poor Country Industrialization," 1972, 54.
[75] "Educating Engineers for World Development," 1975, 33.
[76] Crick, "International Opportunities for Student Engineers," 1970, 18.
[77] Abrams, "The Impact of Antioch Education Through Experience Abroad," 1979, 177.
[78] "Do You Know…?" 1965, xii; Johnston, "World Class," 1996; McCormack, "New Directions," 1966, 373; Wakeland, "International Education for Engineers," 1990.
[79] IIE, *Open Doors 1962*, 1962, 30; IIE, *Open Doors 1963*, 1963, 34; IIE, *Open Doors 1964*, 1964, 36; IIE, *Open Doors 1965*, 1965, 36; IIE, *Open Doors 1966*, 1966, 36; IIE, *Open Doors 1967*, 1967, 34; IIE, *Open Doors 1968*, 1968, 34; IIE, *Open Doors 1969*, 1969, 52; IIE, *Open Doors 1970*, 1970, 52.

students abroad that year.[80] Discussions of obstacles and barriers to participation also repeated familiar patterns. In 1966, for example, a Berkeley administrator noted that engineers were often discouraged from studying abroad due to concerns about degree completion times, a lack of foreign language expertise, and recognition of the high quality of education available within the U.S.[81]

Similar trends continued into the 1970s. While the IIE stopped tracking the number of American students abroad in the early 1970s, by 1977-1978 they reported 801 active study abroad programs in the U.S., but just seventeen were specifically intended for engineering students.[82] Of these, thirteen were operating in Europe and two each in Mexico and Egypt. By 1980-1981, the total number of Americans studying for credit abroad had risen to more than 30,000.[83] However, only nineteen of nearly 800 active study abroad programs were specific to engineering, with fourteen programs in Europe, one each in Mexico, Egypt, Ghana, and India, and one operating worldwide.[84] Mentions of study abroad in publications like *Engineering Education* and *TECHNOS* also remained scarce during this period, even as ASEE more generally expanded its international profile, coverage, and activities.

Once again, such trends can be linked to persistent skepticism about international education among many engineering educators and administrators. In fact, a 1976 National Research Council (NRC) report on "The Role of U.S. Engineering Schools in Development Assistance" indicated that it was of utmost importance for undergraduate students to develop "genuine competence" in traditional engineering disciplines.[85] This distinguished group of authors – which even included some notable advocates for international education – added that "the student's time is almost entirely occupied by studying to acquire a standard body of knowledge."[86] The report also recommended against an earlier suggestion from U.S. AID to develop curricula for a new discipline of "development engineering."[87]

As another factor worth noting, the 1970s where also a time when development assistance was in an important period of reevaluation and reorientation. U.S. policymakers and funding agencies started to look toward projects and programs that were more equitable and bottom-up in character, and their prior emphasis on providing technical assistance and building institutions abroad gradually gave way to other types of technical cooperation, technology assessment, and technology transfer.[88] These shifts were also reflected in the decline of inter-university contracts and the launching of new academic centers and curricula that emphasized emerging and alternate strategies for providing development assistance. Washington University's School of Engineering and Applied Science, for example, emerged as a leader in this movement when it started offering an undergraduate certificate

[80] IIE, *Open Doors 1970*, 1970, 52.
[81] McCormack, "New Directions in Study Abroad," 1966, 370–371.
[82] IIE, *Open Doors 1978-79*, 1979, 77.
[83] IIE, *Open Doors 1981–82*, 1982, 87.
[84] IIE, *Open Doors 1981–82*, 1982, 178–181.
[85] NRC, The Role of U.S. Engineering Schools in Development Assistance, 1976.
[86] NRC, The Role of U.S. Engineering Schools in Development Assistance, 1976, 19.
[87] NRC, The Role of U.S. Engineering Schools in Development Assistance, 1976, 19.
[88] Bernstein, "The U.S. Role in Poor Country Industrialization," 1972, 55; Long, "Designing Engineering Education for Development Needs," 1975, 67–68.

in International Development Studies and graduate certificate in International Development Technology Studies beginning in the late 1960s.[89] Yet as one might expect, such shifts in orientation often reduced or reconfigured the opportunities available for faculty and students to launch or participate in development-oriented international activities.

In summary, those who advocated work and study abroad experiences for engineering students had largely failed to scale up programs and participation in the 1960s and 1970s, even amidst continued increases in the total number of American students abroad, and ongoing conversations about how to best prepare American engineers for work in both developed and developing contexts. As a result, study abroad in many engineering schools remained the province of individual students or small groups, most traveling to developed countries under the aegis of one-off programs or study abroad add-ons. A modest corps of dedicated faculty and administrators – many who themselves had extensive prior international experience and a passion for international education – helped enroll and support the student participants. And while many of these advocates would have agreed with the title of a 1977 paper – "Engineering Students Abroad: You Should Not Have to Ask Why" – they all too often encountered indifference or even resistance to their cause from many engineering students, faculty, and administrators.[90]

The expansion of international education in engineering would require different and more powerful drivers.

PART 2 – GLOBAL EDUCATION FOR ENGINEERS IN THE 1980s AND 1990s

"COMPETING SUCCESSFULLY INTERNATIONALLY": ENGINEERING EDUCATION FOR NATIONAL INTERESTS

As others have discussed at length, concerns over American economic and industrial competitiveness increased markedly during the 1980s and 1990s.[91] What has not been explored in detail, however, is how these national concerns were reflected in new study abroad initiatives in engineering education. In this section, we show how some educators and administrators advocated engineering student involvement in a range of international experiences during this time period. For many, global education was framed as a means to enhancing national economic competitiveness, partially displacing the diplomacy- and development-oriented justifications prevalent during previous decades.

Important early evidence for this trend can be found in a 1982 special issue of ASEE's *Journal of Engineering Education* on "International Concerns." Editorial coordinator George Bugliarello noted in his introductory remarks that there were three possible roles for engineering education as related to the international interests of the U.S. One was "to assist developing countries in acquiring technological skills through engineering education," including for both humanitarian reasons and "enlightened self interest." Another role mentioned by Bugliarello centered on providing engineering

[89]Morgan, "An International Development Technology Center," 1969.
[90]Johnson, "Engineering Students Abroad," 1977.
[91]Downey, "CAD/CAM Saves the Nation?" 1992; Downey, *The Machine in Me*, 1998; Lucena, *Defending the Nation*, 2005.

students with a more sophisticated view of the world's problems and the associated cultural contexts in which they are situated.[92]

Yet even more suggestively, the foremost role for engineering education that Bugliarello emphasized was "the training of personnel to enable U.S. industry to compete more effectively with foreign industry."[93] He also noted two more specific dimensions of this challenge, namely training American engineers for work abroad and revising U.S. engineering curricula to compete against foreign systems of engineering education. His remarks were echoed in subsequent years. In a 1983 conference paper, for example, Carol Ganz of the U.S. National Science Foundation (NSF) pointed to a widespread lack of awareness regarding the extent to which "international activities for U.S. universities and engineering schools are essential to the future of U.S. industry." She added: "[O]ften these programs are perceived solely as development assistance to other countries or 'cultural' activities, without full appreciation of their potential economic benefits for U.S. industry and trade."[94]

Through the 1980s and into the 1990s, Howard Wakeland of the University of Illinois at Urbana Champaign (UIUC) also championed study and work abroad programs in the name of national competitiveness.[95] Wakeland, who was trained in agricultural and civil engineering, had been involved with technical assistance projects abroad since the 1950s.[96] As Associate Dean of the College of Engineering, Wakeland spearheaded many initiatives to expand the number of engineering students studying abroad. He was also instrumental in founding the Engineering Alliance for Global Education (EAGLE) programs, first at UIUC and later spreading to other universities. In a paper published in 1990, Wakeland framed competitiveness as a core justification for such initiatives:

> Why should an engineering educator be concerned about establishing international capability in engineering graduates? Primarily because the engineering educator believes that these graduates are an increasingly important manpower resource in maintaining the technical competitiveness of our society and, in turn, our standard of living.... Global competitiveness has created an increasingly interdependent and international world...Our creative competitiveness through research and development is, however, already under challenge from Japanese and European countries...Clearly those best able to improve U.S. competitiveness are those skilled both technically and culturally.[97]

Under Wakeland's leadership the UIUC engineering department became a leader in international education. In part, he was able to build on foundations established in previous decades. Since the 1960s, for example, UIUC had been working with IAESTE to place engineering students in technical positions abroad.[98] The school's College of Engineering also launched exchange

---

[92] Bugliarello, "International Concerns in Engineering Education," 1982, 266–267.
[93] Bugliarello, "International Concerns in Engineering Education," 1982, 266.
[94] Ganz, "International Activities of Practicing Engineers," 1983, 637.
[95] Wakeland, "Experimental International Programs in Engineering Education," 1987; Wakeland, "A Model International Program for Undergraduate Engineers," 1989; Wakeland, "International Education for Engineers," 1990.
[96] Goodwin and Nacht, *Abroad and Beyond*, 1988, 98.
[97] Wakeland, "International Education for Engineers," 1990, 123.
[98] Wakeland, "International Education for Engineers," 1990, 126.

programs with France and Germany in 1975. These were specifically focused on technical subjects, making them especially notable for the time. Eight to ten UIUC students participated each year. Similar exchanges with the UK were later added.[99] Building on these programs, UIUC expanded its international profile in 1984 with a groundbreaking initiative called the International Minor. It allowed engineering students to gain foreign language skills, experience abroad, and an area studies concentration, all as part of a standard four-year engineering curriculum.

While UIUC's programs were among the most publicized and well documented, other schools were also providing their engineering students with study abroad opportunities during this period. The 1987 *Insider's Guide to Foreign Study* provides basic information about four other study abroad programs for engineers, including Beaver College and Penn State with programs in England, and the University of Connecticut and Michigan State with programs in Germany.[100]

Some engineering educators focused more specifically on Japan as they advocated for international education, reflecting a larger American preoccupation around this time with one of its leading technological and economic competitors.[101] Reflecting a period shift from Cold War military competition to global economic competition, Julian Gresser of the East Asian Consulting Group explained in 1982 that "[t]he technological battle with the Japanese is really an industrial equivalent of the East-West arms race."[102]

Various statistics and program histories also reveal the increasing popularity of Japan as a study and work abroad destination for American students. In 1981, MIT started placing its students in corporate and university internships in Japan, following a period of language and cultural training. Like the International Minor at UIUC, the MIT-Japan program grew largely out of the efforts of one determined individual, a Professor of Political Science named Richard J. Samuels.[103] The MIT-Japan Program expanded from just five participants in 1983-84 to sixteen in 1987-88 and forty-five by 1990-91.[104]

In addition to MIT, Lehigh University, North Carolina State University, and the University of Wisconsin also had Japanese programs or centers for engineering students by the mid-1980s.[105] New inter-university collaborations between the U.S. and Japan also emerged during this period. The aforementioned Engineering Alliance for Global Education (EAGLE) had evolved into a consortium of eleven universities whose goal was to increase international work and study opportunities for science and engineering students in Pacific Rim countries.[106] Such developments help show how perceived competitive threats, especially from Japan, were having a range of impacts on engineering education, including rising interest in the types of "area studies" programs that had for many decades been all but ignored by engineering students.

[99] Wakeland, "A Model International Program for Undergraduate Engineers," 1989, 432.

[100] Leerburger, The Insider's Guide to Foreign Study, 1987.

[101] Coleman and Samuels, "Applied Japanese Studies for Science and Engineering at American Universities," 1986, 209; Ito and Hamada, "East Asian Studies Curriculum Development," 1987, 183.

[102] Quoted in Grayson, "Leadership or Stagnation?" 1983, 356.

[103] Goodwin and Nacht, *Abroad and Beyond*, 1988, 99–100.

[104] Sone, "Consideration for Global Engineering Education Program," 1992, 689.

[105] Coleman and Samuels, "Applied Japanese Studies for Science and Engineering at American Universities," 1986.

[106] Wakeland, "A Model International Program for Undergraduate Engineers," 1989, 431.

Thus, renewed attempts to internationalize engineering education in the 1980s and 1990s were linked to shifting economic and political forces. The increasingly influential ethos of competitiveness was reconceptualized in terms of competition between nations rather than firms, especially as Japan replaced the Soviet Union as a new kind of threat to America. And it was this new understanding of competitiveness that quickly came to overshadow prior justifications for expanding global engineering education, including those related to foreign policy and development assistance goals. The scaling up of both international programs at UIUC and various Japan programs at other American universities during the 1980s and 1990s suggests that the use of competitiveness as a legitimating rationale for the international education of engineers was a successful strategy. Yet in the sections that follow, we turn our attention to some other motivations for expanding the international opportunities available to engineering students.

THE BENEFITS OF GLOBAL EDUCATION: FROM COMPETITIVENESS TO CULTURE, CITIZENSHIP, AND COLLABORATION

A wide range of other justifications were used to support global engineering education in the 1980s and 1990s, including as related to improving various technical and professional skills, expanding student awareness of the effects of culture on technology, and enhancing their role as citizens. Also discussed were some possible benefits to companies and nations resulting from increased international collaboration.

Further, benefits to individual students, especially in terms of their employment prospects and career pathways, were especially prominent in many commentaries from this period. More specifically, advocates variously discussed how communication, foreign language, and problem solving skills acquired while studying or working abroad could enhance an engineer's chance for career success, especially as globalization trends increased the number and scale of multi-national corporate operations.[107] While these kinds of professional benefits were certainly linked to concerns about economic competitiveness, especially to the extent that the success of U.S. industry was linked to the success of the engineers they employed, they also evidence the existence of discourses that were more student-oriented.

Taking the example of improving communication skills among engineers, Johnson made one early and broad argument along these lines in 1977 when he noted that study abroad benefited not only engineers' inter-cultural communication skills, but also their communication abilities more generally.[108] Suresh Chandra and Jane Chandra – the former a Dean of the School of Engineering at North Carolina A&T State University, the latter a German teacher who had herself participated in junior year abroad as a DePauw University student – placed a similar emphasis on communication skills in a 1982 article, including by quoting a 1979 *Newsweek* article that asked, "Aren't we severely

---

[107]Chandra and Chandra, "Junior Year Abroad," 1982; Hartman, "Modern Foreign Languages," 1982; Knepler, "Technology Is Not Enough," 1982; Wakeland, "Trends in Engineering Education," 1985, 1062; Wakeland, "Experimental International Programs in Engineering Education," 1987, 191.
[108]Johnson, Engineering Students Abroad," 1977.

handicapped by the lack of technical people able to communicate on a one-on-one basis, share technology, negotiate, and build goodwill by displaying regard for other languages and cultures?"[109]

Still another alternate justification builds on our preceding discussion of technical assistance and the dynamics of technology exchange. While these issues had emerged in previous decades, during the 1980s and 1990s it was repeatedly emphasized that engineers needed more awareness of the extent to which their work was "culture-bound." As engineering educators recognized a continued lack of awareness among students regarding the challenges of successfully applying U.S. or western technologies abroad, they again pitched cultural studies and study abroad as possible remedies.[110]

During the 1980s, other authors discussed how international experiences could enhance the role of engineering students as members of society and engaged citizens. In 1982, for instance, Sprinkle concluded his article on overseas experiences for engineering students by stating: "Providing at least an initial exposure to other peoples and cultures as part of the engineering student's academic experience can only serve to make future engineers more effective in their roles as influential members of society."[111] Advancing the citizenship argument more explicitly, faculty member Hossein Hakim of Worcester Polytechnic Institute (WPI) justified study abroad programs by arguing that "[s]tudents emerging from these programs are uniquely prepared for citizenship and professional careers in a global economy."[112]

Hakim also went on to develop a more nuanced discussion of the multifaceted benefits of such programs, including many motivations beyond national competitiveness:

> Concerns for national productivity and competitiveness play important roles as well, in leading students, parents, faculty, and program sponsors to recognize the importance of making science and engineering students aware of the global interdependence of technology and commerce. But these should not be the total rationale for our glob- alization effort. The economic considerations which have so mesmerized American commentators are taking place within political and cultural upheavals of at least equal significance. The virtual disappearance of East-West tensions have forced the United States and its allies to acknowledge global dangers which those tensions served to mask, such as environmental deterioration, persistent North-South economic disparities, world poverty, over population and the like…Such exemplary programs will create the kind of technologically-based professional required for continued national and global well-being in the next century.[113]

Such comments echoed the remarks of a handful of engineering educators who had in previous decades discussed international experiences as enhancing the role of engineers as leaders, "world citizens," and even purveyors of "engineering statesmanship."

[109]Chandra and Chandra, "Junior Year Abroad," 1982, 280.

[110]Beedle, "International Perspective of an Engineer's Studies," 1984, 845; Hakim, "Global Perspective Program," 1991, 97; Ito and Hamada, "East Asian Studies Curriculum Development," 1987, 184; Knepler, "Technology Is Not Enough," 1982, 276.

[111]Sprinkle, "Overseas Experiences for Engineering Students," 1982, 285.

[112]Hakim, "Global Perspective Program," 1991, 97.

[113]Hakim, "Global Perspective Program," 1991, 100.

Some programs were also framed as encouraging mutually beneficial technical collaborations. The field of nuclear engineering, for example, provided study abroad opportunities that were described as collaborative.[114] Such initiatives stand in a much longer line of international exchanges in science and technology that have placed more emphasis on generating new knowledge and technologies and less on cross-national competition. A 1987 article by F. Karl Willenbrock, Executive Director of the American Society of Engineering Education (ASEE), similarly stressed that international cooperation remained important alongside competition.[115]

It is further worth noting that much of the rhetoric from Japan surrounding its own exchange initiatives tended to emphasize collaboration over competition. Nissan's Sone, for instance, did not discuss Japan's economic or technological competitiveness among the benefits he saw in global engineering exchange programs. Instead, he indicated that these initiatives could cultivate an international perspective, foster mutual international understandings, and produce engineers capable of working with others from different R&D cultures.[116] Such explanations reflect a long historical tendency for the Japanese to conceptualize progress in more holistic and mutually beneficial terms, especially as compared to the more competitive ethos common in the West.[117]

It is clear, then, that while competitiveness remained a dominant justification for internationalizing engineering education in the 1980s and 1990s, many other motivations persisted. These were sometimes linked to arguments about competitiveness, but not exclusively so. They help reveal the multifaceted nature of global educational experiences including its potential benefits to the United States and its employers, as well as to engineering students as employees, professionals, and citizens. And sometimes, as the following quotation from Wakeland suggests, the various possible benefits to different stakeholders sometimes got intertwined, even if their compatibility with one another might ultimately prove questionable:

> International competitiveness requires cultural awareness, international experience, and language skills…In this age of global technical competitiveness, engineering educators owe it to their students, and to the nation, to help them become not only good engineers but also potential world leaders.[118]

Nontheless, our research suggests that few if any commentators active during this period were addressing the possible tensions latent in such statements.

CHALLENGES AND KEYS FOR SUCCESS: ENABLING GLOBAL ENGINEERING EDUCATION

While a clear enthusiasm and firm belief in the value of study abroad was shared by many of the commentators cited above, it would be inaccurate to paint a picture of extensive support within the

---

[114]Peddicord, "International Collaboration and Exchange in Nuclear Energy Education," 1988.
[115]Willenbrock, "International Cooperation in Engineering," 1987.
[116]Sone, "Consideration for Global Engineering Education Program," 1992, 691–2.
[117]Rosenthal & Matsushita, "Competition in Japan and the West," 1997.
[118]Wakeland, "International Education for Engineers," 1990, 131.

engineering education community as a whole. Despite the growth in study abroad opportunities specifically developed for engineering students, they remained a small fraction of total study abroad programs, and wider enthusiasm for study abroad among engineering educators was often lackluster. Evidence that study abroad was not widely embraced by engineering educators can be found throughout the 1980s and 1990s.

Already in 1982, one finds Robert M. Sprinkle lamenting that "[d]eclining support for international educational concerns is occurring paradoxically at a time when U.S. economic involvement with the rest of the world has never been greater."[119] In 1986, Coleman and Samuels more specifically noted a lack of sympathy for applied Japanese studies among many engineering professors.[120] And after conducting interviews with some American engineering faculty, in 1987 Goodwin and Nacht reported some possible reasons for this widespread lack of an international orientation:

> Most faculty and administrators of professional technical and technological schools with whom we spoke argued that the task of coping with peculiar foreign cultures were not proper problems for their students. Perhaps the responsibility for such matters lay with business managers, lawyers, or area studies specialists. The engineer's function, they said, was to design and to build things, not to worry about the larger culture in which this construction was embedded. Of particular significance to us, of course, was the implication of this argument that there is no purpose in a technical professional either studying about or in a foreign land.[121]

While concerns about competitiveness provided significant impetus for new study abroad programs and related initiatives in the 1980s and 1990's, even this justification failed to turn the majority of engineering educators into advocates for international education.

One can also find other obstacles, including many of rather practical character, which inhibited the internationalization of engineering education during his period. In fact, many of the challenges cited most often in the 1980s and 1990s were similar to those encountered in previous periods, including cost issues and inflexible, overloaded engineering curricula. Chandra and Chandra, echoing a long-standing concern, added that foreign language requirements were a deterrent to engineering students.[122] Additionally, some engineering educators believed that unless study abroad opportunities could be guaranteed in countries where their students would likely work in the future, they were of little value.[123]

Factors that seemed to contribute to successes during this same period also followed familiar patterns. They included having one or more committed, enthusiastic, and visionary change agents, many who themselves possessed extensive international experience. In fact, most of the successful programs described above had at least one faculty member fully devoted to ensuring its continuity success. Sustained institutional support, as mentioned above, also proved to be a vital factor in the

[119] Sprinkle, "Overseas Experiences for Engineering Students," 1982, 284.
[120] Coleman and Samuels, "Applied Japanese Studies for Science and Engineering at American Universities," 1986.
[121] Goodwin and Nacht, *Abroad and Beyond*, 1988, 97.
[122] Chandra and Chandra, "Junior Year Abroad," 1982, 280.
[123] Johnson, "Engineering Students Abroad," 1977, 125.

success or failure of many programs. Although we do not know the stories of most failed efforts to internationalize, Wakeland's personal account of his success provides insights about why others may have failed. He emphasized the importance of having the freedom to develop and experiment with new programs without strict control from administrators.[124] Enthusiasm from individual professors was another necessary but not sufficient condition, especially if higher-level support from their college or university was not forthcoming.

In summary, the increasingly global nature of U.S. business, the end of the Cold War, and Japan's economic ascent all helped establish competitiveness as a leading national priority in the 1980s. Engineering educators responded to these global political and economic shifts, including by working to provide international experiences for their students. Study and work abroad, foreign language courses, and area studies programs all received increased attention. The related argument that engineers with international experience would be one step ahead in the job market also surfaced, as did various claims about enhancing the ability of students to serve as leaders and citizens. On the other hand, discussions about development as a driver for internationalizing engineering education – which had been prominent from the post-war era into the 1970s – were scarce by comparison.

PART 3 – "GLOBALIZING" ENGINEERING EDUCATION: CONTEMPORARY DEVELOPMENTS

As the 1990s progressed, concerns about American competitiveness became intertwined with – and even subsumed by – newly influential discourses about globalization. The founding of a "Global Perspectives Program" at Worcester Polytechnic Institute (WPI) in 1990-1991 provides early evidence for this trend.[125] It was specifically aimed at "globalizing all aspects of the WPI educational program," including through comprehensive plans for faculty and curriculum development, building institutional infrastructures, and providing a wide range of global educational opportunities for students. While ambitious, the initiative built on well-established foundations, including the "WPI Plan," an innovative curriculum established in the late 1960s and early 1970s that required students to undertake real-world projects and engage the social dimensions of their work. WPI also had a longer history of international education, dating back to its first student exchange agreement with London's City College in 1973.[126] By 1991, the school maintained international exchange and project programs in fourteen countries.[127]

By the mid-1990s, globalization themes were becoming even more prominent in discussions about internationalizing engineering education. Russel C. Jones – at the time, a research professor at the University of Delaware and also a former faculty member in MIT's Inter-American Program in Civil Engineering – worked in these directions in 1995.[128] While noting that it was important that engineers understand the dynamics of international competition, he placed greater emphasis

---

[124]Goodwin and Nacht, *Abroad and Beyond*, 1988, 99.
[125]Hakim, "Global Perspective Program," 1991.
[126]Schachterle and Watkins, "The WPI Interactive Qualifying Project," 1992, 52.
[127]Hakim, "Global Perspective Program," 1991, 96–97.
[128]Jones, "Formation of Engineers for International Practice," 1995.

on preparing engineers for "international practice," the "global marketplace," and employment in multinational corporations.[129] Jones added that preparing for such work demanded foreign language proficiency and an enhanced understanding of local cultural, historical, business, professional, and technical issues and dynamics.[130] He also pointed to a handful of schools that were working to "internationalise" their engineering curricula, including Dartmouth, Delaware, Georgia Tech, Union College, University of Rhode Island, and the U.S. Air Force Academy.

Along similar lines, in 1996 Norman Fortenberry acknowledged the growing prominence of America's trade partners, while also emphasizing globalization's manifold impacts on the engineering profession.[131] In addition to being one of the first commentators to use the phrase "global engineering education," Fortenberry called for an integration of "international languages, cultures, and perspectives" into the engineering curricula.[132] He also noted emerging opportunities for student exchange, including through new cooperative agreements between the United States and Europe, and he mentioned Notre Dame as one of "a few pioneering engineering departments in offering a foreign study program for its undergraduates."[133]

Private sector interest in providing global educational experiences for engineers was also evident during this period, especially among multinational corporations. In 1997, for instance, Boeing and Rensselaer Polytechnic Institute (RPI) held an "Engineering Futures Conference." One key outcome of this event was a "Manifesto for Global Engineering Education," which laid out an expansive list of desired attributes for the so-called "global engineer."[134] In addition to a wide range of more traditional professional and technical skills, the report identified attributes such as multidisciplinarity, understanding engineering in context, appreciating other cultures and diversity, working effectively in teams, and upholding high ethical standards.

The establishment of the American-European Education Exchange (AE3) in 1994 created new opportunities for linking existing global engineering programs and lowering barriers to launching new ones.[135] Originally funded by the NSF, Department of Education, and industry, the program – later renamed Global Engineering Education Exchange (GE3) – provided study and internship abroad opportunities for U.S. engineering students.[136] As RPI Associate Dean and GE3 champion Les Gerhardt explained in one report, the initiative could help train "mobile professionals" ready for practice in an increasingly globalized world.[137] By the late 1990s, it claimed seventy university members, twenty-eight in the U.S.[138] By 1998-1999, the number of participating U.S. students had risen to seventy-one, with most travelling to Europe.[139] Yet as Gerhardt noted around this time,

[129] Jones, "Formation of Engineers for International Practice," 1995, 8.
[130] Jones, "Formation of Engineers for International Practice," 1995, 9–15.
[131] Fortenberry, "Under the Jeweler's Loupe," 1996, 169.
[132] Fortenberry, "Under the Jeweler's Loupe," 1996, 171.
[133] Fortenberry, "Under the Jeweler's Loupe," 1996, 169–171.
[134] A Manifesto for Global Engineering Education, 1997.
[135] Gerhardt and Martin, "The Global Engineering Education Exchange Program," 1999, 11b7–10.
[136] Gerhardt and Martin, "The Global Engineering Education Exchange Program," 1999, 11b7–11.
[137] Gerhardt and Martin, "The Global Engineering Education Exchange Program," 1999, 11b7–13.
[138] Gerhardt and Martin, "The Global Engineering Education Exchange Program," 1999, 11b7–10.
[139] Gerhardt and Martin, "The Global Engineering Education Exchange Program," 1999, 11b7–12.

continuing to scale up such exchange programs demanded "total" support from universities and their faculties, along with incentives to encourage U.S. student participation.[140]

While these kinds of initiatives were promising, other evidence reveals that the number of American engineering students traveling abroad increased only modestly from the 1990s into the 2000s. IIE data show that engineering majors ranged from 1.3% to 1.6% of all American students studying abroad for credit between 1985-86 and 1991-92.[141] These figures rose into the range of 1.9% to 2.9% from 1993-1994 to 2002-2003, with an upward trend during this period.[142] For 2002-2003, for instance, about 5,000 engineering students were among the nearly 175,000 American students abroad that year.[143] Parkinson, on the other hand, estimated that more than 5,500 engineering students studied abroad in 2003-2004, representing about 7.5% of that year's graduating class of engineers.[144] According to IIE, more than 6,500 engineering students studied abroad for credit in 2005-2006, increasing to about 7,400 in 2006-2007, or 3.1% of all such students.[145]

These recent upward trends are encouraging, but they also fail to tell the full story. Students traveling abroad may have radically different types of experiences depending on the types of programs they participate in, their destinations, and durations of stay. Parkinson, for example, has identified a wide variety of program formats, including dual degree, exchange, extended field-trip, extension, internship or co-op, mentored travel, partner sub-contracts, project-based/service learning, and research abroad.[146] Still other engineering students participate in rich international experiences without even leaving the United States.

Further, evidence suggests that a large majority of the engineering students who go abroad every year still come from a relatively small number of "pioneering" engineering schools. University of Pittsburgh professor Larry Shuman, for example, estimates that 10 to 15% of U.S. engineering schools are taking international experience seriously.[147] And in a 2007 article, Parkinson profiled global educational programs for engineering students at twenty-three schools.[148] He also noted the participation of about thirty-seven U.S. universities and fifty non-U.S. universities in the afore-mentioned GE3 program, with about 200 U.S. students traveling abroad annually.[149] According to another source, some leading schools have more than a third of their engineering students participating in international experiences. In 2007, for example, Georgia Tech claimed that 34% of its students were studying abroad.[150] And in 2008, RPI announced a plan to have 25% of its engineering students studying or working abroad by 2009 and 100% by 2015.[151]

---

[140] Gerhardt and Martin, "The Global Engineering Education Exchange Program," 1999, 11b7–13.
[141] IIE, *Open Doors 2003*, 2003, 61.
[142] IIE, *Open Doors 2004*, 2004, 61.
[143] IIE, *Open Doors 2004*, 2004, 61.
[144] Parkinson, "Engineering Study Abroad Programs," 2007, 2.
[145] IIE, "Open Doors 2008," 2004b, 1.
[146] Parkinson, "Engineering Study Abroad," 2007, 2–3.
[147] Bremer, "Engineering the World," 2008, 16.
[148] Parkinson, "Engineering Study Abroad Programs," 2007, 5–8.
[149] Parkinson, "Engineering Study Abroad Programs," 2007, 6.
[150] Connell, "Georgia Tech's Well-Engineered Engagement with the World," 2007, 38–46.
[151] "Rensselaer Launches International Experience for All Engineering Students," 2008.

Globalization discourses have persisted in recent years as an influential driver for these kinds of initiatives, what are increasingly referred to using the term "global engineering education." Parkinson, for one, introduced his study by stating that "engineering is a global enterprise," and he went on to note important globalization trends such as the increasing prevalence of corporate outsourcing and multi-national design teams.[152] He also cited NAE President William Wulf, who in 2004 stated:

> [E]ngineering is now practiced in a global holistic business context, and engineers must design under constraints that reflect that context. In the future, understanding other languages, and communicating with other people from marketing and finance will be just as fundamental to the practice of engineering as physics and calculus.[153]

Along similar lines, in 2007 the National Association of International Educators (NAFSA) dedicated an issue of its *International Educator* magazine to the topic of "Engineering and Globalization." As one of the feature articles nicely summarized, "The importance of preparing future engineers to have an international mindset is crucial to their profession as it moves toward a more interculturally collaborative education on a global scale."[154] This same article also noted growing demand from employers since the 1990s for graduates with international experience, making it clear that engineering students might enhance their prospects for career success by becoming more global.

Yet diplomatic, development and humanitarian concerns also resurfaced in the 1990s and 2000s as notable – albeit often secondary – reasons for scaling up global engineering education, including at the undergraduate level. Parkinson, for instance, noted that another reason for providing international experience for engineering students was growing recognition of the "range and scale of technological needs of mankind in the 21st century," including basic needs in developing countries.[155] Along similar lines, Bremer's reference to development-oriented programs like Engineers Without Borders was accompanied by the argument that "[a] global perspective is also required for engineers to tackle worldwide problems."[156]

But perhaps no other recent document better represents the diverse array of rationales for global engineering education than the Newport Declaration.[157] Signed in 2008 by a group of influential stakeholders, the Declaration emphasizes the importance of preparing future engineers for practice in a "contemporary global environment" characterized by rapid social and technical change, while also acknowledging concerns about national competitiveness and national security. But this same statement also notes looming global challenges, such as population growth and resource pressures, and emphasizes the importance of promoting global outreach, community, and collaboration. In fact, the authors go so far as to note how American engineers might serve as "ambassadors" to the

[152] Parkinson, "Engineering Study Abroad Programs," 2007, 1.
[153] Quoted in Parkinson, "Engineering Study Abroad Programs," 2007, 1.
[154] Bremer, "Engineering the World," 2007, 30–37.
[155] Parkinson, "Engineering Study Abroad," 2007, 1.
[156] Bremer, "Engineering the World," 2008, 14.
[157] Grandin and Hirleman, "Educating Engineers as Global Citizens," 2009, 25.

world, and one report in which the Declaration appears was given the suggestive title, *Educating Engineers as World Citizens*.[158]

In an important sense, discussions about internationalizing engineering education have come full circle. The Newport Declaration succinctly captures how globalization dynamics and discourses have created a powerful unifying vision and rationale for global engineering education, including by subsuming other legitimating drivers that have longer historical legacies, ranging from development to competitiveness to diplomacy. Yet, the number of engineering students studying or working abroad continues to rise at a rate much slower than what many promoters envision. In fact, one recent report identified sixteen major obstacles to expanding global engineering education, ranging from rigid curricula and a lack of support from key stakeholders to a shortage of language training and cultural preparation among American students.[159] As the preceding account makes clear, many of these barriers have very long histories indeed. And ultimately, only time will tell whether the types of arguments put forward in documents like the Newport Declaration will help trump these challenges, thereby encouraging a continued and sustainable scale-up of global engineering education.

REFERENCES

A Manifesto for Global Engineering Education: Summary Report of the Engineering Futures Conference, January 22–23, 1997. The Boeing Company and Rensselaer Polytechnic Institute, 1997. DOI: 10.1007/BF01080548 65

Abrams, Irwin. "The Impact of Antioch Education Through Experience Abroad." *Alternative Higher Education* 3, no. 3 (1979): 176–187. 50, 55

Alic, John A., Martha Caldwell, and Robert R. Miller. "The Role of Engineering Education in Industrial Competitiveness." *Engineering Education* 72, no. 4 (1982): 269–273.

Allen, George V. "Where Do We Stand on Point Four?" *Engineering Education: Proceedings of the American Society for Engineering Education* 57 (1949–1950): 65-69. 48, 49

Andrus, J. Russell. "Technical Assistance Through Inter-University Contracts." *Higher Education* XII, no. 5 (January 1956): 75–80. 50

Armsby, Henry H. "An Estimate of Graduate Engineers and Engineering Jobs, 1950–1965." *Journal of Engineering Education* 49 (March 1958): 309-313. 50

Asimow, Morris and John S. McNown. "Engineering Education and International Development." *Engineering Education* 56, no. 3 (November 1965): 65–70. 53, 54

"Backland's Capitalism." *TIME* (August 14, 1964). http://www.time.com/time/magazine/article/0,9171,897285,00.html Accessed April 24, 2009. 53

---

158 Grandin and Hirleman, "Educating Engineers as Global Citizens," 2009.
159 Grandin and Hirleman, "Educating Engineers as Global Citizens," 2009, 13–14.

Barry, Merton. "Foreign Schools Attract Students." *The Wisconsin Engineer* 78, no. 2 (October 1973): 14. 52

Beedle, Lynn S. "International Perspective of an Engineer's Studies." *Civil Engineering for Practicing and Design Engineers* 3 (1984): 843–849. 61

Bennett, Henry G. "The Point 4 Program: A Challenge to Engineering." *Journal of Engineering Education* 59 (November 1951): 139–145. 48

Bernstein, Joel. "The U.S. Role in Poor Country Industrialization: Implications for American Engineering Schools." *TECHNOS* 1, no. 2 (April-June 1972): 49–56. 55, 56

Bibber, Harold W. "The Background of 'Sister University' Contracts Between American and Foreign Universities." *Journal of Engineering Education* 47, no. 3 (November 1956): 196–198. 50

Bjork, Richard E. "Engineering Education in an International Perspective." *Journal of Engineering Education* 50, no. 9 (May 1960): 683–689. 48, 50

Bremer, Darlene. "Engineering the World." *International Educator* XVI, no. 6 (November-December 2007): 30–37. 67

Bremer, Darlene. "Engineering the World." *The Online Journal of Global Engineering Education* 3, no. 2 (2008): 13–18. 66, 67

Bu, Liping. "Educational Exchange and Cultural Diplomacy in the Cold War." *Journal of American Studies*, 33 (1999): 393–415. DOI: 10.1017/S0021875899006167 47

Bugliarello, George. "International Concerns in Engineering Education." *Engineering Education* 72, no. 4 (1982): 266–268. 58

Burr, Anne E. "M.I.T. Involved in Three Major International Programs." *TECHNOS* 1, no. 2 (April-June 1972): 68–69. 53

Chandra, Suresh and Jane Chandra. "Junior Year Abroad: An International Dimension to Engineering Education." *Engineering Education* 72, no. 4 (1982): 280–283. 60, 61, 63

Christensen, N. A. "Social and Political Forces Affecting Civil Engineering Education." *Journal of Engineering Education* 59 (May 1952): 429–431. 49

Coleman, Samuel K. and Richard J. Samuels. "Applied Japanese Studies for Science and Engineering at American Universities." *Engineering Education* 76, no. 4 (1986): 206–210. DOI: 10.1146/annurev.soc.24.1.517 59, 63

Connell, Christopher. "Georgia Tech's Well-Engineered Engagement with the World." *International Educator* XVI, no. 6 (November-December 2007): 38–46. 66

Cooke, Morris L. "The Role of the Engineer in Community Affairs." *Journal of Engineering Education* 41, no. 2 (October 1951): 68–74. 49

Crick, Jeffrey W. "International Opportunities for Student Engineers." *The Wisconsin Engineer* 75, no. 3 (December 1970): 13, 18. 52, 55

Crossley, F. R. E. "Foreign Summer Jobs: A Report on the Operation of the IAESTE." *Journal of Engineering Education* 49, no. 3 (December 1958): 223–227. 51

Downey, Gary Lee. "CAD/CAM Saves the Nation?: Toward an Anthropology of Technology." *Knowledge and Society* 9 (1992): 143–168. 57

Downey, Gary Lee. *The Machine in Me: An Anthropologist Sits Among Engineers*. New York: Routledge, 1998. 57

"Do You Know…?" *Journal of Engineering Education* 51, no. 4 (January 1961): 260. 51

"Do You Know…?" *Journal of Engineering Education* 55, no. 7 (March 1965): xii. 55

"Educating Engineers for World Development (World Congress and Plenary Session Report)." *Engineering Education* 66, no. 1 (October 1975): 33–34. 52, 55

Education and World Affairs. *The University Looks Abroad: Approaches to World Affairs at Six American Universities.* New York, NY: Walker and Company, 1965. 52

Engelmann, Peter. "Engineers in Developing Countries (Conference Reports)" *TECHNOS* 1 (1972): 63–66. 55

"Engineering for International Service," *Journal of Engineering Education* 50, no. 4 (1960): 296. 54

Ferguson, Mark. "75 Years of the Junior Year in Munich." *Die Unterrichtspraxis / Teaching German* 40, no. 2 (Fall 2007) 124–132. 47

Fortenberry, Norman L. "Under the Jeweler's Loupe: Global Engineering Education." *Computer Applications in Engineering Education* 4, no. 2 (1996): 169–172. 65

Ganz, Carole. "International Activities of Practicing Engineers." Paper presented at the ASEE Annual Conference, Rochester, NY, United States, 1983: 633–641. 58

Gerhardt, Lester A. and Shaun Martin. "The Global Engineering Education Exchange Program – A Worldwide Initiative." Paper presented at the 29th Annual ASEE/IEEE Frontiers in Education Conference, San Juan, Puerto Rico, 1999: 11b7–10-13. DOI: 10.1109/FIE.1999.839245 65, 66

Goodwin, Craufurd D. and Michael Nacht. *Abroad and Beyond: Patterns in American Overseas Education*. New York: Cambridge University Press, 1988. 58, 59, 63, 64

Grandin, John and E. Dan Hirleman. "Educating Engineers as Global Citizens: A Call for Action / A Report of the National Summit Meeting on the Globalization of Engineering Education." *Online Journal of Global Engineering Education* 4, no. 1 (2009). 67, 68

Grayson, Lawrence P. "Leadership or Stagnation? A Role for Technology in Mathematics, Science and Engineering Education." *Engineering Education* 73 (1983): 356–366. 59

Hakim, Hossein. "Global Perspective Program: WPI's Response to Global Challenges." Paper presented at the 21st Annual ASEE/IEEE Frontiers in Education Conference, West Lafayette, IN, United States, 1991: 96–100. 61, 64

Hartman, J. Paul. "Modern Foreign Languages: A Needed Requirement in Undergraduate Engineering Education." Paper presented at the 12th Annual ASEE/IEEE Frontiers in Education Conference, Columbia, SC, United States, 1982: 167–170. DOI: 10.1109/FIE.1982.715861 60

Hughlett, Lloyd J. "American Engineering and Industry Abroad." *Mechanical Engineering* 71, no. 7 (July 1949): 571–576. 48

Institute of International Education (IIE). *Open Doors 1955–56.* New York, NY: Institute of International Education, 1956. 47, 50

Institute of International Education (IIE). *Open Doors 1957.* New York, NY: Institute of International Education, 1957. 50

Institute of International Education (IIE). *Open Doors 1961.* New York, NY: Institute of International Education, 1961. 47, 50

Institute of International Education (IIE). *Open Doors 1962.* New York, NY: Institute of International Education, 1962. 55

Institute of International Education (IIE). *Open Doors 1963.* New York, NY: Institute of International Education, 1963. 55

Institute of International Education (IIE). *Open Doors 1964.* New York, NY: Institute of International Education, 1964. 55

Institute of International Education (IIE). *Open Doors 1965.* New York, NY: Institute of International Education, 1965. 55

Institute of International Education (IIE). *Open Doors 1966.* New York, NY: Institute of International Education, 1966. 55

Institute of International Education (IIE). *Open Doors 1967.* New York, NY: Institute of International Education, 1967. 55

Institute of International Education (IIE). *Open Doors 1968.* New York, NY: Institute of International Education, 1968. 55

Institute of International Education (IIE). *Open Doors 1969.* New York, NY: Institute of International Education, 1969. 55

Institute of International Education (IIE). *Open Doors 1970.* New York, NY: Institute of International Education, 1970. 55, 56

Institute of International Education (IIE). *Open Doors 1978–79.* New York, NY: Institute of International Education, 1979. 56

Institute of International Education (IIE). *Open Doors 1981–82.* New York, NY: Institute of International Education, 1982. 56

Institute of International Education (IIE). *Open Doors 2003.* New York, NY: Institute of International Education, 2003. 66

Institute of International Education (IIE). *Open Doors 2004.* New York, NY: Institute of International Education, 2004. 66

Institute of International Education (IIE). "Open Doors 2008 'Fast Facts.'" New York, NY: Institute of International Education, 2004. http://www.opendoors.iienetwork.org/file_depot/0--10000000/0-10000/3390/folder/68485/Open+Doors+Fast+Facts+2008.pdf Accessed April 28, 2009. 66

Institute of International Education (IIE). "A Brief History of IIE." (n.d.) http://www.iie.org/en/Who-We-Are/History Accessed September 5, 2010. 47

Institute of Research on Overseas Programs. *The International Programs of American Universities: An Inventory and Analysis.* East Lansing, MI: Michigan State University, October 1958. 50

Ireson, W. Grant. "Some Problems in Technical Assistance to LDC Universities." *TECHNOS* 4, no. 1 (1975): 13–23. 55

Ito, Barbara D. and Tomoko Hamada. "East Asian Studies Curriculum Development: Preparing Engineers for the Pacific Rim Challenge." Paper presented at the 17th Annual Frontiers in Education Conference, Terra Haute, IN, United States, 1987: 183–187. 59, 61

Johnson, R. Curtis. "Engineering Students Abroad: You Should Not Have to Ask Why." Paper presented at the Second Pacific Chemical Engineering Congress (PAChEC '77), Denver, CO, United States, 1977: 125–127. 57, 60, 63

Johnson, Walter. *1600 Pennsylvania Avenue.* Little, Brown and Company, 1960. 47

Johnston, Theresa. "World Class." *Stanford Magazine* (November-December 1996). `http://www.stanfordalumni.org/news/magazine/1996/novdec/articles/` `worldclass.html` Accessed April 23, 2009. 55

Jones, Russel C. "Formation of Engineers for International Practice." *Australasian Journal of Engineering Education* 6, no. 1 (1995): 7–17. 64, 65

Justice, Howard K. and Josef Wischeidt, Jr. "Pan-American Exchanges in Engineering, Science, and Agriculture." *Journal of Engineering Education* 51, no. 5 (February 1961): 445–459. 50

Knepler, Henry. "Technology is not Enough." *Engineering Education* 72, no. 4 (1982): 274–279. 60, 61

Kusel, Ken. "A Change of Place: International Engineering Programs." *The Wisconsin Engineer* 83, no. 3 (February 1979): 26. 52

Lackner, Jack D., William R. Reilly, and Victor W. GoldSchmidt. "A Latin American Engineering Internship Program." *Engineering Education* 61, no. 3 (December 1970): 295–296. 54

Leerburger, Benedict A. *The Insider's Guide to Foreign Study.* Reading, MA: Addison-Wesley, 1987. 59

Long, Bill W. "Designing Engineering Education for Development Needs." *TECHNOS* 4 (April-June 1975): 65–75. 48, 56

Lucena, Juan C. *Defending the Nation: U.S. Policy making to Create Scientists and Engineers from Sputnik to the 'War against Terrorism'.* Lanham, MD: University of America Press, 2005. 57

Lyman, J., A. Rosenstein, M. Rubinstein, and Wm. D. Van Vorst. "Morris Asimow, Engineering and Applied Science: Los Angeles." In *University of California: In Memorium, 1985*, edited by David Krogh. Oakland, CA: University of California Academic Senate, 1985: 13–15. 53

McCormack, William. "New Directions in Study Abroad: Opportunities for Students in the Professional Schools." *The Journal of Higher Education* 37, no. 7 (October 1966): 369–376. DOI: 10.2307/1979088 55, 56

McGarry, Frederick J. "The M.I.T. Inter-American Program in Civil Engineering." *Engineering Education* 57, no. 2 (October 1966): 111. 52, 53

McKee, Paul B. "The Role of the Engineer in International Affairs." *Journal of Engineering Education* 41, no. 3 (November 1951): 155–160. 48, 49

Merrill, John. "The Scientific Manpower Shortage and Its Implications for the Engineering Technician." *IRE Transactions on Education* 3, no. 1 (February 1960): 3–10. DOI: 10.1109/TE.1960.4322113 50

Miller, Charles L. and Frederick J. McGarry. "Inter-American Program: Past, Present, and Future." *Journal of Professional Activities, Proceedings of the American Society of Civil Engineers* 92, no. 2 (December 1966): 7–14. 52, 53

Miller, Clark A. "'An Effective Instrument of Peace': Scientific Cooperation as an Instrument of U.S. Foreign Policy,1938–1950." *Osiris* 21, Science, Technology, and International Affairs (2006): 133–160. 47

Morgan, Robert P. "An International Development Technology Center." *Engineering Education* 60, no. 3 (November 1969): 247–249. 57

National Planning Association (NPA) Special Policy Committee on Technical Cooperation. "Technical Cooperation – Sowing the Seeds of Progress." Washington, DC: National Planning Association, June 1955. 48

National Planning Association (NPA) Special Policy Committee on Technical Cooperation. "The Role of Universities in Technical Cooperation." Washington, DC: National Planning Association, July 1955. 48

National Research Council (NRC) Board on Science and Technology for International Development, Commission on International Relations. *The Role of U.S. Engineering Schools in Development Assistance.* Washington, DC: National Academy of Sciences – National Academy of Engineering, 1976. 48, 56

Parkinson, Alan. "Engineering Study Abroad Programs: Formats, Challenges, Best Practices." *The Online Journal for Global Engineering Education* 2, no. 2 (2007). 66, 67

Peddicord, K.L. (Session Organizer). "International Collaboration and Exchange in Nuclear Energy Education." Transactions of the American Nuclear Society, International Conference on Nuclear Fission, Washington DC, United States, 1988. 62

Raducha, Joan A. *I'll Remember This Trip': Fifty Years of Study Abroad at UW-Madison.* Madison, WI: Division of International Studies, University of Wisconsin-Madison, 2008. 51, 52

"Rensselaer Launches International Experience for All Engineering Students." Troy, NY: Rensselaer Polytechnic Institute, April 11, 2008. http://news.rpi.edu/update.do?artcenterkey=2422 Accessed April 27, 2009. 66

Rosenthal, Douglas E. and Mitsuo Matsushita. "Competition in Japan and the West: Can the Approaches Be Reconciled?" In *Global Competition Policy,* edited by Edward M. Graham and J. David Richardson. Washington DC: Institute for International Economics, 1997, 313–338. 62

Schachterle, Lance and Maria Watkins. "The WPI Interactive Qualifying Project: A Model for British Engineering Education?" *Engineering Science and Education Journal* 1, no. 1 (February 1992): 49–56. 64

Schwaneger, Henry. "The Junior Year Abroad: Then, Now, and ?" *Die Unterrichtspraxis / Teaching German* 3, no. 1 (Spring 1970): 154–159. 46, 47

Shen, Richard T. "Technical Assistance in Northeast Brazil, 1962–1968 (Discussion)." *TECHNOS* 3 (July-September 1974): 68–71. 54

Shank, Donald J. "Exchange Education in Engineering." *Journal of Engineering Education* 50, no. 4 (January 1960): 328–329. 51

Smith, Craig B. "Overseas Branches of American Universities." *Engineering Education* 59, no. 3 (November 1968): 271–272. 53

Sone, Masazumi. "Consideration for Global Engineering Education Program." Paper presented at the 22nd Annual Frontiers in Education Conference, Nashville, TN, United States, 1992: 688–692. DOI: 10.1109/FIE.1992.683488 59, 62

Sprinkle, Robert M. "Overseas Experiences for Engineering Students: A Brief Primer." *Engineering Education* 72, no. 4 (1982): 284–285. 61, 63

Spychalla, William. "The Wisconsin-Monterrey Exchange Program." *The Wisconsin Engineer* 69, no. 8 (May 1965): 14–19. 52

Steinberg, S. S. "Engineering Know-How for the World." *Journal of Engineering Education* 56 (1948–1949): 411. 49

Thompson, James S. "International Contacts for Engineering Educators." *Engineering Education: Proceedings of the American Society for Engineering Education* 56 (1948–1949): 135–141. 48, 50

Truman, Harry. "Truman Inaugural Address, January 20, 1949." Harry S. Truman Library and Museum (1949). http://www.trumanlibrary.org/whistlestop/50yr_archive/ inagural20jan1949.htm Accessed September 6, 2010. 48

"UCLA Engineering: History." UCLA Henry Samueli School of Engineering and Applied Science (n.d.). http://www.engineer.ucla.edu/history/timeline55_64.html Accessed April 24, 2009. 53

Wakeland, Howard L. "Trends in Engineering Education." *Society of Automotive Engineers Technical Paper Series* (1985). 60

Wakeland, Howard L. "Experimental International Programs in Engineering Education." Paper presented at the 17th Annual Frontiers in Education Annual Conference, Terra Haute, IN, United States, 1987: 191–195. 58, 60

Wakeland, Howard L. "A Model International Program for Undergraduate Engineers." *Engineering Education* 79, no. 3 (1989): 430–433. 58, 59

Wakeland, Howard L. "International Education for Engineers: A Working Model." *Annals of the American Academy of Political and Social Science* Vol. 511 (1990): 122–131. DOI: 10.1177/0002716290511001010 55, 58, 62

Walton, Whitney. "Internationalism and the Junior Year Abroad: American Students in France in the 1920s and 1930s." *Diplomatic History* 29, no. 2 (April 2005): 255–278. 46, 47

Wengler, John. "UW's International Engineering Champion." *The Wisconsin Engineer* 85, no. 4 (May 7, 1981): 7–10. 51

Willenbrock, F. Karl. "International Cooperation in Engineering." *Engineering Education* 78, no. 1 (1987): 12–14. 62

Willard, John A. "Europe Looks to America for Technological Assistance." *Mechanical Engineering* 70, no. 10 (October 1948): 813–814. 48

Woodson, Thomas T. and Murray Mack Gilkeson. "Technical Assistance in Communities in Northeast Brazil, 1962–1968." *TECHNOS* 1 (October-December 1972): 27–37. 53, 54

Zinn, Howard. *A People's History of the United States: 1492-Present.* New York, NY: Harper and Row Publishers, 1980. 49

PART II

Redefining Engineering: Nine Engineers

CHAPTER 3

Crossing Borders: My Journey at WPI

Rick Vaz

RESPONDING TO AN UNEXPECTED OPPORTUNITY

In the early 1990s, I was a recently-tenured faculty member in the Electrical and Computer Engineering Department at my alma mater, Worcester Polytechnic Institute (WPI). My professional focus at the time was on the teaching, research, and service I performed on WPI's compact, fifty-acre urban campus, and my intention was to continue to develop along a conventional faculty career path. As with all WPI faculty members, a key aspect of my teaching duty was to advise undergraduate student projects. Since the 1970s – when I was a WPI undergraduate – every student has been required to complete a series of projects in order to graduate. So, it was no great surprise to me when an undergraduate named Andrew came to my door and asked if I would be willing to advise his senior project. What I did not know was how this project would change my career and life.

Open-ended projects form the centerpiece of the "WPI Plan," an undergraduate curriculum with few required courses, an idiosyncratic calendar and grading system, and an emphasis on learning by doing. In the late 1960s, a group of idealistic young faculty led a revolution that somehow resulted in the replacement of a stodgy curriculum with a project-based approach focusing on learning outcomes and student abilities, and as a student in the late 1970s, I got to experience the WPI Plan firsthand. The junior year project, called the Interactive Qualifying Project (IQP), involves problems at the interface between society and technology or science. This project is intended to help students understand the social and humanistic contexts of their future careers. My interdisciplinary project, which involved mathematics teaching and curriculum revision at a local middle school, certainly influenced my later decision to teach. The Major Qualifying Project is a senior year project in the major field intended to provide an entry-level professional experience; my senior project, building and testing a novel technique for electric motor control, had been a challenging and satisfying experience that helped me land my first engineering job.

The learning objectives of the two projects at the heart of the WPI Plan are centered on solving real problems, and lend themselves readily to working with external organizations. Both projects are completed in small teams – typically two to four students – under faculty supervision, and both require that students take an active role in seeking out faculty advisors and in defining the work and goals. The opportunity to work with students in this very personal and unstructured way had been a major factor in my decision to accept a position at WPI, and I had come to value project advising as a highly rewarding form of teaching. I invited Andrew to tell me more about the project he hoped to complete.

Andrew, a computer engineer, said that he wanted to team up with a computer scientist and a mechanical engineer to design and implement a device to measure tide heights and flow rates with ultrasonic sensors. The application, he said, was intended for the canals of Venice. "Will we get to go to Venice to test it?" I asked jokingly. "Yes," he replied, "there's funding from UNESCO for all of us to go there next summer for installation and testing." Sure enough, nine months later I found myself checking into Boston's Logan Airport with a large, metal enclosure full of circuit boards, tubes, sensors, and other suspicious artifacts, headed to Venice for three weeks. During the academic year, the students had developed the system, and having done some preliminary testing in WPI's swimming pool, they were ready to put the system to the test in the famed waterways of the city known as *la serenissima*, "the most serene." As I sat on the plane, my mood was not one of serenity; I wondered just what I had gotten myself into. Of course, I was looking forward to experiencing Venice and helping the students bring their work to fruition, but I spoke only a smattering of Italian and had little sense of how we would negotiate local authorities and constraints. Arriving in Venice armed with memorized Italian phrases for "I have nothing to declare," and, "We are working for UNESCO to measure the flow rates in the canals," I was, to say the least, a bit apprehensive.

In retrospect, I suppose I should have asked a few more questions of my colleague Fabio Carrera, a native Venetian who had arranged the project and helped us choose a quiet neighborhood for testing since I soon found myself being interrogated by elderly residents and, in one memorable case, armed Italian military police, about why we were drilling holes in the aging canal walls for brackets and fixtures. Somehow, though, it all worked out. As we pursued our work, we were awash in motivation to further our Italian language skills. Every day presented new challenges related to culture and local knowledge, from negotiating purchases in small hardware stores to predicting tidal flows in the labyrinthine canals. We arose early each morning and planned our strategies and goals for the day over typical Italian breakfasts of espresso and pastry. Usually, those strategies had nothing to do with the circuits, software, or sensors, but, instead, involved figuring out how to say and do things in a place where we were decidedly out of our comfort zones.

My students and I were hosted by my WPI colleague, Bland Addison, a French historian who was spending two months in Venice helping Fabio supervise six student teams who were completing their junior year Interactive Qualifying Projects. Shortly after WPI adopted this interdisciplinary project requirement in the 1970s, faculty began establishing off-campus "Project Centers" where students could complete their projects by working full-time on challenges posed by local organizations. The first such center, in Washington, DC, focused on public policy, education, and safety recommendations. The success of the DC center led to the establishment of programs in overseas locations, starting with London in mid 1980s, and followed shortly thereafter by programs in Bangkok and Venice. These international centers are part of WPI's Global Perspective Program, a dynamic network that currently has twenty-six Project Centers in fifteen countries. Although the phrase *Project Center* may suggest a purposely-designed academic space, most WPI centers involve no permanent infrastructure; students and faculty live in rented apartments and complete their work in the offices of their sponsoring agencies and the local settings where their research is conducted.

Fabio had established the WPI Venice Project Center about five years before my initial visit to focus on social and cultural challenges that could be addressed by technology and science, organized around the general mission of helping to ensure a sustainable future for the fragile city and its surrounding lagoon. The projects aligned with his interest in promoting the use of "city knowledge" – data and experience about a city's infrastructure, inhabitants, culture, commerce, and history – to enable more informed decision-making in both public and private sectors. The projects that summer involved teams of WPI students working with the City of Venice and other local organizations to preserve public art, monitor impacts of boat traffic, and recommend sustainable solutions for preserving the fragile lagoon ecosystem.

I was fascinated by the range of issues involved in these interdisciplinary projects, and I was struck by how they challenged the students to think critically, communicate effectively, and understand the social and cultural contexts of the problems they addressed. Students concerned with restoration of public art, for example, had to consider not just restoration techniques but also the city's priorities and political processes, artisans' motivations and traditions, residents' concerns, and a host of local statutes and regulations. This type of contextual thinking had been envisioned by the WPI faculty who had developed the projects-based WPI Plan in the late 1960s. Influenced by the social changes of the decade, this group had set out to develop a curriculum that would produce "technological humanists," cognizant of the social impact of their professional work and committed to solving problems in ways that were both practical and principled.

The WPI Plan was highly unusual at its inception, and it remains remarkable today. The belief that project work should be at the heart of an undergraduate education led to a program with few required courses, no prerequisites, no failing grades, and an emphasis on teamwork and hands-on learning. Throughout the evolution of the Plan, the Interactive Qualifying Project has remained at the heart of this vision; student teams are formed from every major on campus, and faculty from every department and discipline – engineering, physical science, social science, management, and the humanities and arts – are expected to engage in advising real-world, interdisciplinary research projects.

Bland and Fabio certainly seemed up to the challenge. I admired the way they wove in their respective viewpoints – those of the historian and the urbanologist – to guide the students' work by illuminating it through different disciplinary lenses. I was envious of my colleagues' opportunity to spend two months challenging and mentoring student researchers working on a range of interesting and practical issues while immersing themselves in a rich and fascinating culture. I found Venice irresistible, but in a way, that transcended my touristic interests; Venice, by virtue of its complexity, was a wonderful laboratory for students to think about the cultural and social contexts of their work. My two colleagues enthusiastically welcomed my interest in the work that they and their students were pursuing, so that in addition to supervising my own students, I was able to get involved in the other projects as well. I resolved to learn more about this rich opportunity for teaching and learning when I returned to campus.

MAKING AN INTENTIONAL COMMITMENT

After my return, I pursued the opportunity to advise Interactive Qualifying Projects at an off-campus center, offering to go wherever I was needed. Within six months, I received an assignment to WPI's Washington, DC Project Center and was able to accept the assignment thanks to the support of a department head who saw the value of such experiences. Project advising was and is viewed as a central aspect of teaching at WPI, but assignments to advise away from campus for a full seven-week term require flexibility on the part of everyone involved. At this time, only about 10% of WPI students completed off-campus projects, so just a handful of advisors were needed, and there was no formal process for involving faculty. WPI faculty normally teach four courses per year, meaning that an off-campus assignment usually displaces one course in a faculty member's schedule. To accommodate the assignment, the Provost's Office provides funding to the faculty member's home department. Still, such assignments can disrupt the teaching, research, and service work of a department, and those assignments are only approved if the department head concurs. My department head, who had completed a similar assignment himself when he was a junior faculty member, was an active supporter of the program, resulting in many faculty members from the Electrical Engineering Department becoming involved over time. Although the program now enjoys the support of department heads across campus, at the time, I was fortunate to be encouraged to take on what was still an unusual assignment for a WPI faculty member.

Susan Vernon-Gerstenfeld, the social scientist who directed the Washington, DC Project Center, also taught the preparation course required of the DC-bound students, a sort of "research boot camp." Sue was an excellent mentor to the students and to me, as I learned to help the teams frame achievable goals, develop focused literature reviews, choose research methods, and integrate all of that work into a set of tightly written proposals. I worked with fifteen students organized into five project teams, working for sponsors such as the National Science Foundation, the US Coast Guard, and the Consumer Product Safety Commission. In each case, the technical or scientific issues of the project were relatively straightforward compared to the challenges posed by political, social, and human factors. Compared to my own engineering research, or that of my graduate students, this work involved far more nuances – different schools of thought, conflicting expert opinions, and a multiplicity of options for collecting and analyzing data.

The experience in DC was intense and rewarding, as I found myself having to play a wide range of roles that were new to me. If a student was depressed or anxious, I was challenged to act as a counselor. When homesickness arose, I found myself in a parenting mode. When negotiating project work with sponsors, I was the sole representative of the university, and sometimes had to assert the primacy of WPI's academic goals over the sponsors' shorter-term interest in data collection. I quickly had to come up to speed on methods for qualitative research and analysis. Perhaps the most challenging situations, though, arose from team dynamics, as the teams of students worked closely together for fifty hours per week and often encountered differences in work style, commitment, and opinion. I found myself relishing the challenge of so many new roles, as the assignment called upon me to use and refine a much wider set of skills and abilities than I was used to using back on campus. Only occasionally did I feel overwhelmed. Most of the time, I felt like a diplomat, a therapist, a sociologist, an entrepreneur, and a mentor.

The students, most of whom had never been asked to work so hard, also seemed to find the experience exhilarating. Most students exhibited significant personal growth, especially in the form of increased confidence and a more mature attitude; I came to attribute this to being taken well out of their comfort zones, both in terms of surroundings and disciplinary content. Much of the growth, I imagine, arose from the fact that for the first time in their educational careers, they were being asked to do something that really mattered to someone else. The sponsors were highly enthusiastic about the students' work, from which they saw real benefits, and supported the students on a daily basis, in some cases, developing strong mentoring relationships with them. The culminating oral presentations at the sponsoring agencies were polished and professional thanks to hours of practice, and the students' pride in their accomplishments was evident. So was mine – as an educator. I was hooked on this new form of teaching and learning.

Upon returning to campus from Washington, I felt compelled to tell others about my experience and to find ways to keep it alive. I regaled any colleagues who would stop to listen with my enthusiastic stories about the unexpected challenges, refreshing change of pace from campus life, and affirming evidence of student growth. It was like returning from a short but very satisfying sabbatical. I found when I was back in Worcester that I missed the intensity of being at the Project Center. Although my duties on campus were rewarding and demanding, I had found it exhilarating to adapt to new surroundings, rush to attend meetings across the city, and to attend to unexpected problems and student concerns on my own. I slowly adjusted back to my "normal" existence, teaching courses, writing papers, and enduring committee meetings – but I also made the decision to advise off-campus on a yearly basis. In retrospect, I think I craved the unpredictability of working with students far from home. I was soon pleased to accept the assignment to advise the following year at a relatively new Project Center in the city of Delft, the Netherlands – part of WPI's rapidly expanding network of international sites.

My experience in Washington had taught me the basics of running an off-campus operation and guiding interdisciplinary student work. Two two-month stints advising in Delft helped me begin to see the added educational impact of international experiences. The Netherlands is not, on the whole, a difficult place for US citizens to visit – English is widely spoken, for example, and public transportation is efficient and simple to navigate. However, our interactions with the Dutch provided some fascinating lessons for my students and me to ponder regarding differences in cultural perspectives and worldview, challenging many of our assumptions, and introducing us to new ways of thinking. In particular, the notion of sustainability—still a largely unknown term in the US during the mid 1990s—wove through the project work in ways that caused my students and me to think about science and engineering in broader ways than we had before. The projects in Delft involved issues such as urban planning, infrastructure, energy, and economic development, and I was struck by how the theme of sustainability wove its way through aspects of each project. It reflected a national sensibility that, as an engineer, I found irresistible – the Dutch were actually trying to make decisions that would have multiple long-term benefits, and balance the imperatives of prosperity, the environment, and quality of life.

I also began to understand some of the complexities of negotiating cultural difference, and the value to students of such understanding. Students studying urban design, for example, had to first understand the roles of private and public space in Dutch culture, which differ greatly from those in the US, and how those roles inform appropriate designs of green space. Students recommending sustainable energy plans soon understood that the views of Europeans regarding energy use, environmental impact, and the role of government were not the same as those of their home culture.

With each off-campus assignment, I became more interested in students' personal and intellectual growth, especially in helping them think about the cultural and social contexts of their projects. I was struck by the fact that we learned at least as much about ourselves and the United States as we did about the host country and its culture. I came to the conclusion that my students were so much a product of US culture that they required some distance between them and home to think about culture with some objectivity. I developed strategies for providing students with formative feedback, not just on their writing, but also on their critical thinking, their teamwork, their professionalism, and their interactions with the sponsors and with me. I also took time to get to know the students and to discuss their career goals, their personal aspirations, and reflections about their experiences off-campus. Most were eager to join me for a meal to discuss matters beyond their project work. I was finding these assignments to be the most powerful teaching experiences of my career, leading to wonderful and diverse opportunities for my students' intellectual and personal growth.

RESPONDING TO A NEW PASSION

During the first part of my career, I had approached engineering mostly as an interesting intellectual challenge. However, due to the project work I was now advising, particularly in areas related to civil engineering, the profession began to take on new meaning for me as I learned more about how technological solutions shape and are shaped by social forces and human needs. I found myself paying a lot more attention to interrelated aspects of energy, the environment, civic infrastructure, and quality of life. Energized, I chose to become more involved in the organizational aspects of the Global Perspective Program; by the late 1990s, I was serving as the Center Director of WPI's Project Centers in Venice and the Netherlands, as well as a smaller program involving senior design projects in Limerick, Ireland.

Being a Center Director appealed to my interest in designing holistic learning experiences, as I became involved in everything from project development and selection of housing to cultural orientation and language training. At this point, my role as an educator had transcended engineering because the focus of my work was development of students more generally as critical thinkers and problem solvers. I also attended to logistical and administrative matters that impact the learning environment. As director, I was the primary liaison between the university and our contacts in each location: housing providers, universities, project sponsors, and, especially, Local Coordinators. Most Project Centers have a Local Coordinator on retainer to help the Director make contacts, solve housing and transportation challenges, negotiate cultural minefields, and plan cultural events. These Coordinators tend to be individuals with some tie to WPI – an alumnus, for example, or the parent of a graduate – and a long term commitment to the success of the program and the quality of experience for the students. Together, the Local Coordinator and Center Director develop the local research program and design the experience that the students and faculty advisors will have.

In addition to taking an expanded role in program development, I also began to read and write more about the type of educational experiences that students had off-campus. I began participating in engineering education conferences, and found them very stimulating – filled with interesting people looking to share ideas and learn from each other. The time I had once devoted to graduate teaching and disciplinary research gradually gave way to more focus on undergraduate education and the scholarship of teaching and learning, and the presentations and publications I had once given about signal and image processing were soon replaced with talks and papers about project-based learning and internationalizing engineering education. I became increasingly more interested in and concerned with general education learning outcomes. The transformation felt right for me, and also right for WPI, since I saw leadership opportunities for us as other engineering programs began to experiment with project-based learning and international experiences. I found sources of affirmation for my new perspectives and interests as I began building relationships with like-minded colleagues from around the nation.

In the late 1990s, the engineering education community was starting to take a more active interest in internationalizing and broadening the curriculum, and my work overseas had developed into a passion that aligned well with those emerging interests. The Accreditation Board for Engineering and Technology's Engineering Criteria 2000 were already on the horizon and beginning to spur curricular changes at many institutions, although few at the time had a clear vision of how to meet all of the criteria. Around the same time, WPI had brought in a new academic administration with a stated commitment to internationalizing the campus. The Global Perspective Program had grown considerably, with Project Centers in locations including Bangkok, London, Costa Rica, Puerto Rico, Denmark, and Zurich, and almost half of all WPI students were now completing off-campus projects, with about 30% of those having an international experience. In 1998, I was asked by WPI's Provost to assume a half-time administrative role to support the rapidly growing Global Perspective Program. My work focused specifically on faculty and student development in preparation for project work abroad. I began to immerse myself in the detailed workings of international education, and I became increasingly involved in discussions about engineering education reform, both on campus and nationally.

I could sense that some of my engineering colleagues were puzzled by my involvement in these international activities, and I was aware that a portion of the WPI faculty was skeptical of the value of international project work for engineers and scientists. Usually, this skepticism was based on concerns that broad, interdisciplinary learning experiences would come at the expense of disciplinary depth; concerns that will exist as long as universities are organized into disciplinary groups. However, the senior faculty from across campus who had begun off-campus project programs in the 1970s and 1980s provided a strong and supportive mentoring network within which I and others were able to grow and learn, and over the coming years, I saw many of the skeptics gradually replaced by newer colleagues, many of whom have been eager to get involved in the program. At the same time, the program was itself transforming, as both my colleagues and our students, in increasing numbers, were seeking experiences that would take us further out of our comfort zones.

VENTURING FURTHER AFIELD

In 1999, I traveled with a colleague and thirty students to advise projects in Thailand. The notion of living and working in a developing country in Southeast Asia was both enticing and intimidating. With the benefit of thoughtful mentoring and guidance from Steve Weininger and John Zeugner, two colleagues with considerable experience in Thailand, I prepared myself to undertake a new type of teaching and learning. The projects in Thailand included community-based research, involving a strong component of service and presenting significant challenges for cultural preparation. My students, mostly from the middle-class towns and suburbs of New England, soon found themselves working to address basic needs of underserved communities, both in the slums of Bangkok and the villages of rural Thailand. We worked mostly with non-governmental organizations and not-for-profit organizations on projects related to air and water quality, economic opportunity, education, public health, and agriculture.

None of my advising experiences in Europe or the US had prepared me adequately to work in Thailand. The friendly populace, fabulous food, and amazing sights that attract visitors from around the world were the tip of a cultural iceberg that was rooted in a collectivist, high-context culture that regularly confounds and enlightens the thoughtful visitor. Although Thai culture and language preparation were important components of the students' preparation, the amount of culture and language they could absorb in one semester was limited to survival skills. Hence, learning and thinking about culture became a central part of our work in Thailand.

My students struggled to develop solutions that would be locally appropriate and sustainable in communities with many challenges and few resources. Often, a key component of the projects involved interviews, focus groups, or surveys of populations that spoke no English and employed very indirect communication styles. Even with the assistance of translators eager to improve their English, the students regularly saw their careful plans disintegrate on site, to be replaced by more modest and realistic approaches to the research. The students learned the importance of building trusting relationships, of framing problems based on community perspectives, and of drawing on local knowledge to solve those problems. They also learned the necessity of flexibility, adaptability, and creativity, three attributes essential to both engineering practice and international field research. Confronted with the complexities and contradictions of international development, the students' cultural experiences caused them to rethink many of the assumptions they held about technological advance, social justice, and a sustainable future. Their work caused them to consider the views of multiple stakeholders and to wrestle with competing economic, environmental, and social issues; as a result, their simplistic assumptions about how the world works were often replaced with more nuanced and complex perspectives.

In the following years, I returned regularly to Thailand, lured by the chance to help students negotiate their work and learning there. Student interest in the developing world had also led WPI to open Project Centers in Latin America and Sub-Saharan Africa, and I traveled to those locations for advising experiences, each with its unique lessons and cultural revelations. In Puerto Rico, the students worked on reforestation projects, mangrove sustainability, and recycling initiatives. In Namibia, the focus was on water resource management, renewable energy in rural areas, and erosion control in informal settlements. With each trip, I became more convinced of the urgent need for the next generation of engineers and scientists to become aware of fundamental problems related to development and to understand how the solutions to those problems must be consistent with local traditions, resources, and political realities.

I also became convinced that lessons drawn from this type of work could change students' view of the world and what they might do with their lives. I was struck by the powerful learning that students can experience when given the chance to work on problems of importance to underserved communities. In each cohort of students, my colleagues and I would see a handful of students whose life and career goals appeared to be changing before our eyes. Students who had previously envisioned traditional engineering careers applied for Peace Corps assignments, or went to graduate school for public policy, or pursued teaching. Other students, it seemed to me, would "pack away" the experience as one might return a suitcase to the attic, perhaps not to be revisited until some later time. Overall, my colleagues and I became increasingly convinced that student learning and growth opportunities were greater when students were brought further out of their comfort zones, and when we were very intentional in facilitating cultural learning. On the WPI campus, returning students sought ways to build on their experience, pursuing work with sustainable energy or environmental studies and starting student organizations focused on sustainability. Many students who had completed the program became active ambassadors for the experience and for WPI, giving presentations to other students, to Trustees and other visiting dignitaries, and to audiences at professional conferences. Some of these students, it seemed, exhibited an urgent need to talk about their experiences and reflect on the meaning of what they had done.

LOOKING AT WHERE WE ARE NOW

Fifteen years after that first project in Venice, I am now fully immersed in international education. As Dean of WPI's Interdisciplinary and Global Studies Division, I have two primary areas of oversight: the Global Perspective Program and a group of faculty focused on sustainability studies in support of the IQP – a wonderful group of social scientists who are both appreciative and tolerant of engineers. I never expected to find myself serving as a fulltime academic administrator, and I had viewed the former dean's impending retirement with trepidation. I knew that as dean I would have little time for classroom teaching, and I knew that I would miss that aspect of my work with students. I also knew that the move would limit my interactions with the Electrical and Computer Engineering Department, which had been a second home, as student and faculty member, for more than half my life. Finally, I was apprehensive about taking on a role where much of my time would be spent worrying about budgets, addressing personnel challenges, and dealing with unhappy or upset students and parents. However, when the position was offered to me, I decided to accept. The opportunity to facilitate the further growth and development of the Global Perspective Program was irresistible, considering its impact on students and importance to the university as a whole. I was also drawn by the chance to work closely with the faculty and staff who have made the program successful, as well as the opportunity to support the professional development of faculty from across the campus.

After three years, I can say that I enjoy the job far more than I ever expected to. Working with students has always been my favorite part of academic life, and I made sure to continue doing so. I still co-direct our Project Centers in Thailand and Ireland, and regularly advise student work both at those locations and at others. I am fascinated by the so-called "millennial generation," and for the most part find them very well suited to the challenges our program presents. They understand the importance of working in teams, they are exceptionally proficient at using technology to solve problems creatively, and they are driven to make a difference to and help those less fortunate than themselves. On the other hand, they are sometimes uncomfortable with open-ended assignments, they often lack the attention span demanded by careful reading and evidence-based writing, and their helicopter parents are apt to hover over them – either figuratively, with daily phone calls and text messages, or literally, with visits to Project Center sites that can sometimes get in the way of work. Overall, though, I find this generation's spirit and energy very promising. Many of them already seem to understand that they are citizens of the world – they are concerned about the planet's most pressing problems, such as energy, water, climate change, and public health.

The opportunity to work with faculty from different disciplines across campus is also reward-ing; I enjoy the support of a cohort of experienced and dedicated colleagues who have built the program over the years, and we are all excited to see a steady stream of younger faculty getting involved. Since the IQP and Global Program draw on all of WPI's academic departments, I make it my business to get to know new colleagues and present opportunities for them to get involved if they wish. Increasingly, the Global Program is being seen as a way to attract talented faculty to WPI; I am often asked to meet with promising faculty candidates to explain the opportunities they will have to work with students on compelling interdisciplinary problems around the globe. Once new faculty arrive on campus, my colleagues and I participate in faculty orientation, involve them in workshops, present invitations to co-advise projects, and try to develop an understanding of each new colleague's interests in international education. Of course, involving newly arrived, tenure-track faculty in international activity requires judgment and caution. If a faculty member's department head is uncomfortable with the assignment, or if there is reason to believe that his or her tenure case requires that full attention be paid to scholarly work, that person's involvement is in nobody's best interest at that time. In my experience, though, off-campus advising provides a wonderful professional development opportunity for newer colleagues to learn how to direct and mentor student researchers.

Another rewarding aspect of my job has been involvement in the various academic communities that have organized themselves around international engineering education and other aspects of educational reform. I have learned much, entered some fruitful collaborations, and forged lasting friendships. In 2004, I had the opportunity to reflect on my experiences in engineering education during a sabbatical leave at the Association of American Colleges and Universities (AAC&U), a time I devoted to writing projects and campus initiatives centered on internationalization and reform, especially for science, technology, engineering, and mathematics (STEM) disciplines, but also for other disciplines. Recently, the international education and liberal education communities have been paying an increasing amount of attention to engineering, as changing accreditation criteria begin to make possible the vision of engineering as liberal learning to prepare students for the 21st century. In effect, the three communities are coming to convergence around a common set of themes – the importance of cross-cultural abilities, the impact of experiential learning, and the need for programs that could be scaled up to serve most if not all students. This convergence is exciting to me. The experience of working with AAC&U and interacting with colleagues from around the country, in addition to opening up collaborations, has also helped me to appreciate what a remarkable series of opportunities I have enjoyed at WPI, from initial involvement in an established international program all the way to assuming responsibility for what is arguably the university's signature program.

About half of all WPI students—over 350 students per year—now complete at least one academic project outside the United States, and about 30% of WPI faculty members participate in the Global Perspective Program at some stage of their careers. Over thirty years after the first WPI Project Center was established, the program has proven to be both sustainable and scalable. On the WPI campus, student and faculty involvement overseas has become part of the institutional culture, celebrated by the president and trustees and supported by the provost and department heads. Nationally, organizations such as the National Science Foundation and the National Academy of Engineering actively promote global competence for scientists and engineers as a national priority, as do ABET and employers. There is an impressive menu of initiatives to internationalize engineering curricula at campuses across the country and an ever-growing list of workshops, conferences, colloquia, journals, and initiatives centered on international engineering education. It is very exciting to be part of a fundamental shift in engineering education.

LOOKING AHEAD

Much work remains before a substantial proportion of US engineering students will have international experiences. Colleges and universities will have to wrestle with curricula, faculty and campus cultures, and resource issues to internationalize their engineering programs in ways that are scalable and sustainable. WPI has had the advantage of a history of educational innovations and is an institution where undergraduate education is widely viewed as the university's primary mission. Furthermore, we are fortunate both in terms of institutional resources and our student body. Still, the program took years to become established, and today sustained growth of WPI's global activity competes for faculty time with increasing expectations for scholarly productivity, funded research, and a range of other on-campus educational initiatives, including interdisciplinary first-year seminars on issues of global importance. The challenge to develop international programs at other institutions, especially those with high research expectations and resource shortages, is great. What will it take to meet the challenge? In my experience, four key factors in successful internationalization are educational impact, program evaluation, faculty involvement, and sustainable funding.

Those planning international programs should start with a clear understanding of what they are trying to achieve educationally. The learning outcomes of an impactful international program influence all aspects of program design and implementation. An overseas experience, while providing a powerful learning opportunity, does not automatically follow through on that promise. WPI very intentionally chose to use its international programs as a platform for project-based learning, and the success we have enjoyed is based on the critical thinking and problem solving that are emphasized in those projects. Nationwide, while some programs focus on exchange, experiential education is emerging as a very popular model in engineering education abroad, and for good reason – the learning outcomes from experiential education align well with ABET criteria and institutional program outcomes. Experiential education can involve project work, service learning, internships, or other forms, depending on the goals, structure, and duration. Unlike students of the humanities, whose primary focus for study abroad is often on language, history, culture, and other location-specific studies, engineering students may benefit more from learning how to "get things done" in an unfamiliar location than from an intensive study of a specific location. Experiential models, which are of increasing interest in the education abroad community, should be of particular interest to engineering educators, since the educational outcomes are likely to be useful in meeting overall program objectives—and more importantly, in helping students to better understand how to be effective agents in an increasingly interconnected world.

Secondly, a program's educational goals, once established, should be clearly articulated to stakeholders and systematically evaluated. At WPI, I have found that assessment of student learning at our project centers provides persuasive evidence to faculty and administrators regarding the value of the program. Students and their parents, faculty participants, service providers, international partners, and study abroad professionals each may harbor very different assumptions about and expectations for an international experience. Clarifying the educational outcomes to these audiences can help align the efforts and expectations of each group, increasing the likelihood of meeting program goals. Program assessment should be structured around those goals, and I should include both formative and summative aspects. Formative assessment of students' learning and experiences can help shape the way they think about their experiences and also allows programs to respond to challenges and unanticipated needs. Summative assessment can provide key evidence of program impact and value to persuade students, parents, funders, accreditors, administrators, and even skeptical faculty colleagues. Some study abroad outcomes such as cross-cultural competency can be challenging and time-consuming to assess, and require a careful and well-informed assessment design, possibly involving longitudinal assessment. Experiential programs can provide evidence of learning outcomes such as research skills, problem solving, teamwork, and effective communication, so program evaluators should think broadly when considering assessment targets and strategies. Evidence of such outcomes can go a long way toward persuading key audiences on your campus, such as engineering faculty, of the value of an international program for their students.

Thirdly, broad involvement and support from engineering faculty is a key to institutionalizing an international engineering program. The involvement of faculty from the engineering departments across WPI in the Global Perspective Program has been instrumental in achieving greater participation rates among our students. Engineering faculty are role models who directly and indirectly influence their students' choices, whether actively through advising and mentoring, or indirectly through curricular design and the activities they choose to pursue. It is perhaps not surprising that the energy for international engineering programs often comes from study abroad offices and faculty in the humanities and arts. However, without substantive involvement from engineering faculty, programs are unlikely to be well-integrated into the engineering curriculum or seen by most students as important pre-professional experiences. Engineering faculty who are involved in the development and delivery of international programs send their students a clear message about the role of engineers in the world. Although there are many reasons why faculty might or might not get involved, the most important factor is probably the extent to which that involvement is recognized and rewarded by the institution. If internationalization is an institutional goal, then faculty contributions to that goal must be positively reinforced within the systems for recognition, reward, and the allocation of institutional resources.

Finally, the long-term viability of any program depends on a sustainable financial model. Both costs and revenues associated with international programs vary to a great extent based on the type and structure of the program. Typically, program costs are split between participants and the institution, and they can be offset by a variety of revenue streams. The push to internationalize engineering education has resulted in new sources of funding for starting programs and collaborations; the trick, of course, is to structure the activity to be affordable after the funding runs out. WPI's programs were built intentionally over many years, and they have been supported through the operating budget in recognition of their central role in our students' educational programs, and, at the same time, supplemented by grants, partnerships with project sponsors, and gifts. Endowments are ideal for sustaining program and student expenses; corporations and foundations with global agendas are promising targets to endow international activity, as are alumni interested in promoting international experiences. Such resources should be judiciously applied to balance the affordability of programs to students and to the institution, and to enable faculty involvement which, although expensive, is sometimes the best approach to ensuring positive outcomes and promoting student participation.

I am heartened by the growing recognition from leaders both within and outside academia of the need for globally competent engineers and scientists to tackle the world's most pressing problems. Not so long ago, the globalization of engineering education was driven almost exclusively by a handful of committed faculty members championing programs that were totally dependent on their individual involvement. Increasingly, though, programs are being institutionally supported and embraced. Still, I am not convinced that those calling for global competence fully understand the cultural change that will be needed on our campuses, especially those where faculty activity is centered more on funded research than on undergraduate education. I am struck by my good fortune to have been involved with a program that was relatively well-established and that has been able to serve as a national model, and I am aware that its growth and success have been products of WPI's particular circumstances. Facilitating transformative international experiences for so many students and faculty is neither simple nor inexpensive, but it is energizing and rewarding, and seems more important than ever in light of world events. The essence of engineering is problem solving, and the next generation of engineers will face problems related to food, water, energy, the environment, and public health throughout the world. To do so, they will need more than traditional engineering knowledge. They will need firsthand understanding of the local and global contexts in which those problems are embedded – and engineering faculty mentors who have a global perspective as well.

CHAPTER 4

Education of Global Engineers and Global Citizens

E. Dan Hirleman

BACKGROUND

During my years as an educator, I have come to understand there are almost as many ways for students to learn as there are students. A strength of the U.S. system of higher education is the diversity of faculty and staff, instruction styles, courses, and co-curricular activities. Our goal is to ensure each and every student is able to "connect" his or her passion for learning with at least one course, project, program, or member of the faculty/staff. But since the courses and programs are created by faculty and projects are guided by faculty - at the end of the day it is about people-to-people connections. Many of us attribute our passion and success to one or two teachers who took an interest in us as unique individuals and encouraged us to excel. My personal path has all these elements. There were a few special people who inspired me in ways that have framed my efforts in educating global engineering professionals.

A number of colleagues and I at Purdue are working toward a shared goal – we want all engineering students to have a compelling experience in a program, course, project, or a student organization that activates a self-sustaining desire to fit their engineering education into the greater global context. In fact, we are convinced that engineering educators must consider global perspectives as an integral part of a holistic education, along with technical and professional dimensions of the engineering profession.[1] We have made some progress, though there is much more to do. In 1999-2000 there was only one student in the School of Mechanical Engineering at Purdue who had an organized educational or professional experience abroad, and thirty-six in all of the College of Engineering (Schools at Purdue equate to Departments at many universities). In 2009-10 we had eighty-four ME students involved in substantial international experiences related to engineering (nearly 30% of the bachelor's degrees awarded in that year), and over 232 from the College of Engineering.

Global engineering education offers students a remarkable opportunity to lead more expansive lives, both personally and professionally. But, as educators, we must rise to the challenge of implementing a wide range of global engineering education opportunities. We cannot expect one program

[1] Hirleman et al., "The Three Axes of Engineering Education," 2007.

to meet the global engineering needs of an entire college. The importance of multiple pathways has become increasingly evident to me throughout my continued involvement in the field of global engineering education. We have developed pre-freshman year programs, short-term courses abroad in cross-cultural communications, global design team experiences, and our flagship multi-year program (GEARE) that combines twelve credit hours of language and culture, two engineering internships, a semester abroad, and two semesters of global design team project work. Our portfolio of global experiential learning pathways has something to meet the needs of most of our students.

And the impact on the students is quite gratifying as well. Students matriculating through the GEARE and other programs have joined the Peace Corps, taken two years to teach English in Mongolia, joined international MBA programs, and are leading global design teams. They have built global networks of friends and colleagues. One, in his first year out of school, has committed to endowing a $25k scholarship over five years to support students for experiences like he had. They attribute their expanded worldview to the experience Purdue enabled.

OBSERVING THE WORLD (FROM MIDDLE AMERICA)

I can trace the beginnings of my interest in international activities to the 1970s, when I was enrolled in a Thermo II course taught by Professor Sigmar Wittig at Purdue. Because I had done well on the first two exams, and not so well on the homework, Professor Wittig asked if he could speak to me after class. At that time he had been on the faculty at Purdue a few years and was well aware of the differences between the German and U.S. higher education systems. In Germany and most of Europe, all that matters is the final exam. In the U.S., essentially, every assignment and every mid-term counts as well. So I was operating more like a German than an American student in the class, and he took a personal interest. And I was quite intrigued myself, given that my Midwest upbringing had not brought me in contact with many internationals. That personal connection triggered relationships and events that essentially have defined my career ever since.

Originally from Aachen University in Germany, Sigmar Wittig employed the German hierarchical approach to building a research group - undergraduate assistants worked with junior graduate students, and those students then worked under senior graduate students. But he also had diverged from the formal Professor-Dr. style in the sense that he engaged on a personal level with everyone in the group. He and his wife Elizabeth (also from Germany) hosted great social events at their place, all the students looked forward to those days. And we even had a chance to experience a Porsche 911 in action against the 427 Corvette Stingray that the senior graduate student in the group drove.

I had the opportunity to participate in three semesters of undergraduate research, with some flexibility and responsibility in defining the direction. So, as a junior, I already understood the relationship of coursework to discovery activity, and my own personal responsibility for defining the problem, setting a goal, and generating useful results. My work was part of a paper presented by the senior Ph.D. student at the Annual Meeting of the American Physical Society.[2] At the conference, I was really taken by the fact that my engineering work could be, and in fact was, original enough

---

[2] Zallen et al., "Determination of Contact-Region Gradients and Their Influence on Shock Tube Properties," 1972.

to be of interest to people from around the world. I realized for the first time that I could know as much about a topic as anyone, anywhere in the world, and that there were smart people all around the world who did great science and engineering. This sparked one of the driving forces in my professional life – a desire for impact and recognition on a worldwide scale. And it started me on a career trajectory where I worked to create international opportunities for myself and others.

VENTURING INTO THE WORLD (AS AN AMERICAN ENGINEERING STUDENT)

Research and its potential for international presence and impact were now in my blood. I continued at Purdue as a Master's student, working in Professor Wittig's lab on a project related to my undergraduate research. I was intrigued by the very different life experience my international colleagues had and actively pursued opportunities to participate in international engineering communities (which was not a hard sell given the international flavor of my research group). I wrote a paper with my advisor and represented our group at the International Shock Tube Symposium in Kyoto, Japan. I presented a paper on my Master's thesis work,[3] plus another paper for a Ph.D. student in our group who was unable to attend. This was another adrenaline high, at least with regard to international engineering. I was in Japan, presenting papers to a large audience, where Americans were definitely a minority, and having to defend my work. One particularly vivid memory was being congratulated by an English scientist whom I did not know when I answered a question from a Professor (who happened to be from Switzerland) in defense of the other student's paper. This added another quantum of confidence, since I understood someone else's Ph.D. work well enough to defend it under fire in an international setting.

I pursued an international research assignment at the time between my Master's and Doctoral work. With help from Sigmar Wittig I was able to join the lab of a Norwegian (Professor Bjorn Qvale) who had previously been on the faculty at Purdue and was now Director of the *Laboratoriet for Energiteknik* at the Technical University of Denmark (DTH) near Copenhagen. I have always liked adventures, and this was one. I instinctively knew that I would grow in ways that were not possible staying in the Midwest. This also presented my first chance to convince funding agencies and sponsors of the importance of international experience as an engineering student. I needed the support of the National Science Foundation, relative to my NSF Graduate Fellowship (no problem) and Hughes Aircraft Co., relative to a Howard Hughes Doctoral Fellowship (a harder sell). But the sponsors agreed, and I was able to spend more than a year in Denmark and had a chance to integrate to a small degree into another culture.

I lived in student housing among the Danes, interacted (in broken Danish) with the great people in the machine shop, and took time off from working long enough to participate in the essentially-mandatory coffee breaks with all the faculty and staff of the Lab. The student housing was co-ed (not common at Purdue in the 1970s). We took turns cooking, and we all ate together

[3] Hirleman, "Chemi-Ionization in Homogeneous Methane Oxidation," 1974; Hirleman and Wittig, "Microwave Measurements of Non-Equilibrium Electron Densities in Reacting Gas Mixtures," 1975.

in the shared kitchen. I quickly learned how to cook a few "American" dishes for my turns at the "mad klub" (food or meal club), and one of my first calls back home to Kansas was to ask my mother how to cook a baked potato. To make friends it was necessary to spend hours watching the 1974 World Cup (though I did actually begin to enjoy the game the rest of the world calls football), and defend the American version of football. This was my first real cross-cultural experience - I was the minority culture and was forced to look at American culture from the outside.

One of my "kitchen-mates" was a follower of Trotsky, and I was able to hear (and argue for many late nights) about a different perspective on the success and/or failings of the U.S., capitalism, communism, and Christianity. It was enlightening to read Trotskyist newsletters describing a dismal life in America (that did not match my personal experience) and the impending collapse of the U.S. and capitalism. He believed it all, and knew that I believed my orthogonal view as well. It took quite a while to peel back the layers of assumptions in both our world views - in doing so it was always my expectation we would find common ground. We never did, and intellectually parted ways when he said that were he in charge and true to his beliefs, the death penalty would have to be in effect for those who continued to practice Christianity. This was, by far, the starkest articulation of human differences I had personally experienced, and a situation where "not wrong, just different" did not apply in my mind. Since much of the rationale of this opposing worldview was related to the economic plight of the non-rich, the experience reinforced my belief that economic freedom (as classically defined in the U.S.) and the resulting gradual rise in quality of life from economic activity is crucial to mitigating dictatorships of all forms and improving the human lot. I believe the practice of mechanical engineering, via harnessing nature's energy forms, is one way we can empower people and have a positive impact. Certainly it is true that technology developed by engineers can be hijacked for evil purposes, but I think history shows that, in general, poverty breeds miserable leaders and development breeds improved quality of life for the masses.

Another close circle of Danish friends I made and have to this day were the members of the DTU basketball club. I naturally looked for social activities that I was familiar with, as most internationals do, at least according to my observations. I was also sought out, since it is assumed in Denmark that all Americans are good at basketball. Fortunately I am reasonably athletic and had played more hours of pickup basketball than most Danes, so I actually was good enough to play with the team. We participated in some international tournaments, and coupled with travels related to my research interests, I was able to visit much of Western Europe. My ethnic origins are northern European, so I also could blend into a Danish group. I remember several situations where I consciously worked to not stand out as an American, but rather to be a Dane, situations where it was a (albeit minor) disadvantage to be American. And this gave me more windows into ways the rest of the world views Americans.

I also became involved with students from IAESTE (The International Association for the Exchange of Students for Technical Experience), including some from the Eastern Bloc nations as well as Asia. And knowing these students who actually lived in the "communist" world was also a very enriching experience, in that it was clear that many did not believe a word—unlike my Trotsky-follower friend.

Another significant event from this experience in international research involved working to acquire some research equipment. For my research project, and for what turned out to be the subject of my Ph.D. dissertation, I was developing laser-based methods for measuring particle sizes in engine exhausts fueled by methanol-gasoline blends. The technical challenges included processing high frequency signals from the photomultiplier tube detectors that were collecting the scattered light as the smoke particles flew through the interrogation laser beam. I needed to record and analyze signals from several detectors simultaneously, which was a challenge in the mid 1970s. I recall that, through considerable detective work, I ascertained that there were two (Tektronix) oscilloscopes in the entire nation of Denmark at that time that could have done the job. One was at a university in Aarhus, and the other at the Danish Nuclear Energy Research Lab (RISØ) in Rosklide, about 100 km away from the Technical University where I worked. In an act of kindness which still strikes me as remarkable to this day, Dr. Lars Lading of RISØ agreed to let me borrow their 100 MHz, 4-channel storage oscilloscope for July and part of August. I drove to Roskilde, picked it up, brought it back to DTH, and used it to acquire the data I needed for that portion of my project. Certainly my mentor at DTH was known to the other lab, but I do not remember nearly as much paperwork as would be required today or even at that time in the U.S. This reinforced the fact that research engineers and scientists are really an international community with many shared goals, and I am motivated to "pass on" the favor.

The Danish view of the world was very different than that of Middle America, and I recall my friends invoking the saying, "Americans live to work and Danes work to live." I endeavor to have it both ways. I recall being amazed at the exam and course structure in Denmark: no homework, only a final exam. Some friends seemed to do essentially no work for the year spent most of their time in sports and at the University *Kælderbaren* or Cellar Bar (another cultural difference, alcohol on campus) and only started studying just before the exam. If they failed, they had to retake the course the next year, but they were able to stay in the dorm and the university for another year, no problem. One of my dorm mates had even graduated but did not like the career, came back to the university/dorm (with essentially no tuition and highly subsidized housing) and started over in her second major.

By the end of my graduate student days I had been able to present papers in several continents, had friends from around the world, and had been able to look at my culture somewhat from the outside. The combination of all these experiences and relationships was crucial to the direction I have taken
since that time.

CRAFTING AN INTERNATIONAL IDENTITY (AS AN ENGINEER/SCIENTIST/EDUCATOR)

During my early years as a professor at Arizona State University, I chose to engineer my career based on the premise that there is little use competing only regionally – and that success should be measured by recognition and impact on an international scale. Consequently, I pursued activities in international professional societies. A highlight was serving as organizer and chair for the 2nd International Congress on Optical Particle Sizing held in Tempe, Arizona in 1990. Held every third year, this Congress cycles through the Americas, Europe, and Asia.

Serving on an organizing committee with mostly international members provides many more opportunities to experience different values and perspectives. Participation in the conference was highly international, and it was a chance to show a bit of America and an American university to the attendees. As conference chair there are many opportunities that involve interaction with people from very different cultures. The activities include reviewing and accepting/rejecting papers, finding ways to provide travel support to colleagues from countries in the developing world, and dealing with large variations in the level of technical work from various labs. It forces one to become very conscious of the advantages we have in terms of facilities, equipment, and support for research here in the U.S., and to ponder what type of research I would be doing if I had been born into one of those other communities.

A highlight was a post-Congress three-day hike I led deep into the Grand Canyon. There were nearly thirty attendees and friends who went on the hike down the Havasupai Trail (nearly 2 km vertical drop and 13 km to the campground, 18 km to the Colorado River). I will always cherish this memory: each morning began with an opera-singing French researcher serenading the camp, and each day ended with the younger hikers struggling to keep up with Henk van de Hulst, who was one of the fathers of light-scattering research and well past retirement age at the time. Another enjoyable outcome of this event was the article in an issue of *Applied Optics* dedicated to the Congress that also included photographs of the hikers at the top of the Hualapai Hilltop, before and after the hike (a.k.a. the "fried-look" and "refried-look" pictures).[4] It is due to the many personal and professional rewards such as this that I have always encouraged my graduate students to participate in international meetings and societies.

In the early 1990s my family and I took a year's sabbatical leave in Holland and Germany. It was quite rewarding to see our children begin to have international experiences, comparable to those that my wife Laura and I had. But they were having them at a younger, i.e., middle-school, age. Our visit was hosted by Professor Brian Scarlett at the Technical University of Delft and Professor Sigmar Wittig at *Universität* Karlsruhe. This was another chance to experience different cultures, and for all of us to make friends from the local communities and from around the world.

For me that experience and the extended time in Europe continued a process and allowed me to became active in ISO (International Standards Organization) in addition to ASTM (American Society of Testing Materials). Some think that standards committee meetings are tedious and maybe

[4]Hirleman and Bohren, "Optical Particle Sizing," 1991.

even boring. If that is true for those within one culture (e.g., in ASTM), then it is many times more so at the international level where the cultures and their various protocols must be handled effectively to get anything done. This was another great learning experience in cultural differences and attitudes. This too was a considerable effort, but ended up being a success. I was able to chair a group that carried out an international inter-laboratory testing program that provided the support for an international (ISO) standard for calibration of laser diffraction particle sizing instruments. Again, it was clear to me at the time that if my research and innovations were a valuable contribution for the Americas, they must also be valuable to the rest of the world and I then have an obligation to help them be deployed worldwide. Implicit here was deployment in the developed or near-developed world, without much thought of the possibility that not all technology is appropriate for the developing world. Regardless, developing an international identity through exposure to much of the world made the transition from using my gifts and talents

in helping our nation to helping on a global scale an easy one.

GIVING FORWARD - HELPING THE NEXT GENERATION BECOME GLOBAL ENGINEERS

By 1998, I had served about eight years in department and then college-level administrative positions at ASU, and found this quite fulfilling. It was fulfilling because I was effective as a leader, and therefore had much broader influence through helping the faculty reach their own potential. That year I applied to be Head of my alma mater, the Purdue School of Mechanical Engineering. Bringing my experience full circle, as part of the process I submitted a vision statement highlighting my desire to establish strategic partnerships with a few universities around the world that could enhance student interaction around engineering topics. I remain convinced that the international experiences I had early in my career have played a major role in my ability to work and get results at an international scale. Yet I knew that I was definitely in the minority for engineering students - only a very small percentage take advantage of international opportunities for work or study during their undergraduate years. And with rapid expansion in the flow of information and people, international collaboration will become increasingly important. It is important both to leverage resources and also to increase understanding and cooperation around shared concerns since local activities have worldwide impact, for example, in terms of energy and environment.

I moved to Purdue as department head and, along with three other initiatives, started working on globalizing our program in a substantial way. My first attempt at a large-scale program did not succeed. I had developed a friendship and working relationship with Dr. Thomas Malott, an alumnus of the ME school and, at the time, CEO of Siemens Energy and Automation. Tom is from a small community near Purdue, and through education, experiences, and determination worked his way up from a regional engineer to a top position in the world's largest enterprise. Tom had seen firsthand the level of competition for advancement at global corporations like Siemens, and shared with me a visceral concern about globalization, global competition, and preparing engineering graduates accordingly. Clearly, the world of engineering and technology was rapidly changing, and, in fact, continuing improvement and even maintenance of the U.S. standard of living would be in jeopardy if we continued business as usual. Our first attempt to jumpstart a global engineering program in the School of Mechanical Engineering was to seek several million dollars in funding for an endowed joint professorship between Purdue and Karlsruhe. Tom helped me pursue funding from a potential donor, but the request was denied. Nonetheless it was clear to me we were on the right track and providing opportunities of substance for our students was imperative.

In addition to the joint Professorship (which still has not come to fruition), I initiated a dialog with Sigmar Wittig about other ways we could connect our two universities to enrich the educational experiences for students from both institutions. I am convinced that a personal global experience of the right type is the best possible mechanism for an engineer to become a global engineer. And maybe more importantly, a *global engineer* is an engineer who is adept at incorporating requirements from multiple cultures into the selection, definition, and solution of engineering problems. A *global engineer* can go against an instinctive decision appropriate for their own culture with ease, by seeing the point-of-view of another culture that is more appropriate for the context, and make a counter-intuitive decision accordingly. Experiential global engineering education involves *making decisions* in a cross-cultural environment.

Tom and I shared the view that this type of professional, cross-cultural experience is quite different than what is normally considered a study-abroad experience. We wanted to give Purdue students this opportunity. These conversations helped incubate our concept for the Global Engineering Alliance for Research and Education (GEARE).

It was during this time frame that I first met Christian Callegari and Peter Olfs of Siemens. Christian had leadership responsibilities for global university relations at Siemens, and our proposal for a named professorship had found its way to his desk. He needed to understand what these Purdue people were talking about, so I went to Munich to meet Christian and Peter in person and discuss what Purdue was trying to do in terms of educating global engineers. That trip started a very fruitful relationship between Purdue and Siemens that grew to include Karlsruhe and Shanghai Jiao Tong. A large number of Purdue and Karlsruhe students have interned with Siemens and
some have taken permanent positions with the firm.

CREATING *THE* COMPREHENSIVE PROGRAM FOR EDUCATING GLOBAL ENGINEERS - A DESIGN PROBLEM

Building a substantial and sustainable global engineering program at a university is a challenge. There are only two real examples of success deriving from the 1980s, notably Worcester Polytechnic Institute (WPI) and the University of Rhode Island (the International Engineering Program or IEP). These are amazing achievements, and the drivers of these (John Grandin at URI and a group of people including Rick Vaz at WPI) deserve continuing accolades. There are, however, some unique features of those institutions and programs that make them difficult to replicate at a place like Purdue. A couple of key differences are a greater emphasis on research at Purdue and the larger scale (about ten times more students overall and about three times more engineering students). In addition, we at Purdue elected to make *global team* experiences foundational.

For a university educational program to be sustainable, a critical mass of the faculty (well beyond the founder) must engage with and "own" the program. At Purdue that means research and professional connections must be a part of the equation or faculty would not integrate it as part of their core mission. So we concluded that the first criterion of any sustainable program at Purdue is that it must have a research component, the "R" in GEARE. That meant our global partner institutions must be comparable in research reputation and scale to Purdue such that collaboration on research and scholarship among their faculty and ours was compelling.

The second imperative for a global engineering program is that it must represent a compelling value proposition to engineering students. One reason for the relatively low numbers of engineering students participating in substantial global experiences is that Study Abroad and International Program officers often do not understand how to make these programs a compelling value for engineering students. Of course one compelling value proposition is, "do this and then you can graduate," i.e., make it a requirement). To a certain degree that is what WPI has done, greatly helped by the fact that they designate an entire quarter of academic work to the capstone project. That makes it very easy for students to fit a global project with time abroad into their curriculum. But no university to my knowledge has yet required a global (abroad) experience for graduation, and I suspect never will. It is conceivable that private universities could do so, but even the WPI program does not "require" it.

I became very aware of this cultural mismatch between International Programs and engineering at Purdue as we tried to build programs to grow international credit-bearing experiences for our engineering students (from thirty-six of 8,000 or so students enrolled, and of 1,700 or so degrees per year in 1999-2000). Traditional study abroad activities are, in effect, peripheral to an engineering education, epitomized by the example of students cramming all of their technical courses into seven semesters so they can free up a study abroad semester where they can take general studies courses readily available overseas.

Now, a case could be made that engineering students should be happy to take an extra semester or so to graduate if it will expand their horizons and broaden their education. That is roughly the point-of-view normally taken by staff from university-level International Programs Offices since generally they deal with students from majors where that is more the case. The reality is that students with that admirable attitude toward education do not normally find their way into engineering programs and the engineering profession. Furthermore, even interested students cannot always afford an international experience.

Another common misunderstanding with non-engineering colleagues across campus is related to summer activities. In the School of Mechanical Engineering at Purdue our senior exit surveys show that 80-90% of our students have had three or more months of paid engineering internship experience before they graduate with a BSME. This experience provides a crucial credential and substantial pay (typically 75% of a starting salary) to engineering students. Engineers are, in general, quite focused, and they factor all of this into their plans. The equation is quite different for a foreign language major who is considering taking a summer for an international experience, e.g., studying Mandarin Chinese and studying Chinese history in Shanghai. Since there is generally no comparable internship opportunity, and the experience is really a necessity for the language major, it is a "no brainer." For engineering students it can still be compelling, but not nearly so much so as for other majors where the global is recognized as an integral part of being educated.

So to address the issue of having only about 1% of our graduating engineering students involved in formal education abroad, we first tried to understand the "customer." In Mechanical Engineering, we have an introductory one-hour-per-week sophomore level course, now called the Global Professional Seminar. This course is led by Dr. Dianne Atkinson of Purdue, a great contributing partner in our efforts to globalize our program and curriculum. Dr. Atkinson surveyed all students in that course for several semesters and asked about the barriers to studying abroad. The students mention three factors: 1) *cost*, 2) *delay in graduation*, and 3) lack of a *community* or support structure. The first is really more myth than reality - often the total cost of a semester of coursework abroad is about the same as one at home, depending on the country and financial aid specific to study abroad. The second, of increased time-to-graduation, is also more myth, though it is true that advance planning for sequencing coursework is required. The *community* piece involves moving to a new place, leaving family and friends, and often having difficulties in communication.

The GEARE program, or more precisely the GEARE Global Engineering Program for undergraduate students[5] was purposefully designed to mitigate all three of the barriers identified by the students. The design objectives we adopted for the program were the following: (1) cost neutrality, through scholarships and paid internships, (2) no delay in graduation due to the program, i.e., a full load of courses in the necessary prerequisite chain can be taken abroad, and 3) development of a community of typically five to ten students from Purdue and an equal number from our reciprocal partner universities who work together on co-located (face-to-face) design team projects. To add to the value proposition we also included internship experiences with our partner companies, one domestic and one international.

One of the more controversial aspects of our design of the GEARE Global Design program involved language requirements. This is a feature of the program where we took a pragmatic approach to the design problem. Recall that our goal was for 20% of our ME students (and eventually all engineering students) to be involved in GEARE Global Design. To address the graduation time barrier, we had to focus on a very few global strategic partners. We have built partnerships for geographic regions representing about one billion people each. Clearly, for global engineering China had to be involved. A challenge with China is the difficulty of the language. In fact, it is commonly accepted that there are four classes of languages as characterized by the immersion time required to reach different levels of fluency. Mandarin Chinese is in the group of most difficult languages, and estimates are that it takes two years of essentially 100% immersion to reach the highest (level

[5]Chiu et al., "The Purdue Global Engineering Alliance for Research and Education (GEARE) Program," 2007; Groll and Hirleman, Undergraduate GEARE Program," 2007.

4) competency. For an engineer working in China something less than that may be required to be effective. But it was clear the chances of getting substantial numbers of Purdue students, who are predominately from middle-America, to obtain one-and-a-half to two years of 100% immersion in Mandarin was not practical (unless we took native speakers, which is not the goal of the program). So given the choice of relaxing the level of fluency requirement or abandoning China in the GEARE program, it was no contest. We set a requirement of twelve credit hours of language and culture for the target location, which clearly means most or all of our students in China function in English. In contrast, Spanish is among the easiest languages to learn, and, therefore, many of our GEARE students in Mexico operate in Spanish.

Certainly this choice has its critics, often among those who teach language at the university-level. At Purdue, the faculty of Foreign Languages and Engineering have joined forces to develop the GEARE program, and they have found that some students go back for additional language courses after they return. The GEARE program has generated a new group of students who are taking courses in the Foreign Languages and Literature (FLL) Department, which is a win for all. Our colleagues from FLL at Purdue have played a crucial role in developing our global program. For example, Professor Paul Dixon, the head of the department, has served on the Global Engineering Programs Team and has supported our collaboration from the beginning. Professors Beate Allert and Wei Hong have been instrumental in recruiting and interviewing students, offering courses on campus (including working individually with our students for their specific needs), co-teaching courses abroad, and participating in orientation activities.

The GEARE Global Engineer Program has been a success in terms of time to graduation. Of participants to date, 63% complete their degrees in four years, 94% in five years, and 100% in six years. Indeed, GEARE students have better time-to-graduation than the overall engineering student body. The five-year data include three students in the cooperative education program (co-op) who graduated on schedule in that five-year program. In addition, another handful received minors in languages, so even these data really understate the case.

The level of commitment required of students in the GEARE program is quite high. For that reason, we embarked on a path to formally recognize this comprehensive experience, which includes coursework, internships, and global design team projects. We explored both a certificate and a minor label for the experience, querying our industry sponsors, GEARE students, and faculty as to their preference. The consensus was that a minor implies more substance than a certificate, and Global Engineering Studies Minor was chosen for this formal recognition. The intention was not to exclude students who participated in other global activities of equivalent substance, but rather to set the bar such that the Minor carried significant meaning. The first version did require an international internship experience, which we presumed to be an internship in industry.

BROADENING THE GLOBAL LEARNING PORTFOLIO IN M.E.

Our original goal was to have 20% participation by ME students in the GEARE Global Design program. We are currently well below half that target, which is frustrating at times. Students must commit to the program as freshman or early their sophomore year and also to twelve credit hours of language and culture coursework before they go abroad during their junior year. So there is substantial risk and uncertainty involved when the decision must be made so early. In addition, students are encountering increasing opportunities for extra-curricular activities, some of which can preclude participation in the GEARE program.

GEARE supports undergraduate engineering students in a twenty-four-month program that integrates domestic and international internships, study abroad, language and culture orientation, and a two-semester, multinational-design-team project into a comprehensive experience. But this level of immersion is more than many students are willing to take on. For that reason, we became convinced it is necessary to broaden the portfolio of experiences to include some that require a much lower level of commitment, and we set out to deliberately create multiple pathways.

For example, Dr. Atkinson from ME and Professors Beate Allert and Wei Hong from the Foreign Languages and Literature Department developed a two-week course titled, "Introduction to Intercultural Teamwork" that has been offered in Germany and China. While the time duration is limited, the faculty have done an excellent job of making a real immersion experience via considerable interaction with students from the global partner universities (U. Karlsruhe and Shanghai Jiao Tong University), including team projects. Dr. Yating Chang, assistant director of Purdue's Global Engineering Program, along with Carolyn Percifield developed a pre-freshman course on Creativity and Innovation offered in the summer in Singapore. These short-term courses allow students to get involved in a gradual way and, in some cases, help us recruit students to longer-term global experiences. Our faculty have also developed and offered more standard study abroad courses, such as Energy in a Global Context taught by Purdue ME faculty in Germany.

In 2009, we launched GEARE Junior Year for Mechanical Engineering students whereby they spend spring semester of their third year at Shanghai Jiao Tong University. This is designed to impose a lower barrier-to-entry to students not in the 20% who gravitate to GEARE and programs requiring a similar level of commitment. The participants travel as a group, have courses as a group, and have campus housing arranged for them. The courses they take are essentially from the SJTU curriculum and offered by SJTU faculty, with some modifications that allow these course offerings to directly substitute for Purdue courses and thereby satisfy ABET requirements for their Purdue degrees. Students pay their normal Purdue tuition for the semester. In addition, they pay for airfare, food, housing, and other expenses in Shanghai, which in total turns out to be a little less than they would spend living a student lifestyle at Purdue. We have thirteen students in our first offering, with plans for thirty per year in future years.

BROADENING TO INCLUDE OTHER DISCIPLINES – THE GLOBAL ENGINEERING PROGRAM (GEP)

The need for globalized engineering education is clearly not limited to mechanical engineering. The college leadership team made a decision to create the Global Engineering Program (GEP), and I offered to be the interim director during its first couple of years, in part because the ME School had already invested in developing a substantial portfolio, and it was generally applicable to all schools and disciplines. We assembled an interdisciplinary team of committed faculty and staff, known as the Global Engineering Program Team. The team developed a strategic plan, which included a vision for Purdue being "preeminent in educating engineers for global professional competence and in global collaborations for research, education, and engagement." The mission included "synthesizing an integrated portfolio of opportunities, programs, and partnerships that support Purdue students and faculty to be leaders in the global network of engineering professionals."

The GEP supported expansion of the GEARE Global Engineer program to other disciplines, and currently students from Aero/Astro, Civil, Electrical, Civil, and Computer Engineering have participated. In addition, the faculty from other schools have developed courses (e.g., Earthquake Engineering in Turkey offered by Civil Engineering and Zero-Gravity Experimental Methods offered by Aero/Astro Engineering using a Drop Tower in Germany), which also attracted students from many disciplines across the College of Engineering. A conscious effort was made to publicize the opportunities and needs. For example, the Purdue chapter of the GEARE student organization conceived, developed, and operated the Global Opportunities Fair on campus to highlight the variety of opportunities there are for global education.

The GEP experience highlighted the fact that Purdue's ME School has been unique with respect to global engineering education. It has a sizable group of committed faculty, including Professors Fisher, Groll, Krousgill, Meckl, Chiu, and Sojka, who have spent sabbaticals at GEARE partner universities or in target countries. Much of the growth in the College educational programs has derived from this group and ME faculty. One disappointment is the fact that most of our sister departments have not embraced this mission as quickly and to the degree that ME has. I think that speaks to the complexity of developing such programs, and the need for passionate souls who will adopt this as part of their educational and service mission. But there are encouraging signs—GEP has been institutionalized, has a permanent

Director, and has seen an increase to three-and-a-half staff.

RETHINKING ECONOMIC COMPETITIVENESS AS DRIVING RATIONALE – GLOBAL SERVICE LEARNING

The GEARE program was designed with the expectation students would seek employment in industry. This was both because internships are important aspects of our co-curricular learning vision in ME at Purdue and because the opportunity for an international internship with a leading global firm is compelling to students. In the process of developing and getting approval for a Minor in Global Engineering Studies, it became clear that a requirement of global professional practice in industry was too restrictive. An increasing number of students in the current generation are interested in serving humanity directly rather than through developing products or services via a commercial enterprise. Based on feedback from colleagues involved in curriculum decisions across the College of Engineering, Purdue's Global Engineering Studies Minor requires an engineering-profession-related experience abroad of three months or more, and global service learning activities or other engineering-related professional practice experiences satisfy the intent and the requirements.

Many different types of projects can be carried out in the context of global service learning. Several organizations including Engineers Without Borders (EWB), and Engineering for World Health (EWH), among others, are dedicated to these types of activities. Engineers for a Sustainable World (ESW) focuses solely on sustainability, both education and projects, with domestic and some international efforts. Normally these groups and activities involve extended stays at a site abroad where the project is deployed. This type of activity provides students with opportunities to practice a different form of engineering than do the industry internship experiences. Both are valuable and draw different types of students. There seems to be a prevailing sense among students today that to really "help people" or "make a difference," it is necessary to practice a form of humanitarian engineering. While that view is understandable, I think all engineers actually help people, and if we were better at communicating that, we would be more successful in attracting young people to careers in engineering.

One senior capstone project I advised involved designing a water system for Humatas de Anasco, a community in Puerto Rico. This project was inspired by colleagues at the University of Puerto Rico – Mayaguez, Professors Ephraim O'Neill and Luisa Seijo. UPRM had been operating community service learning for many years, and I was pleased to engage with them in a collaborative project. I worked with a faculty member from Purdue CE and Professors O'Neill and Seijo to put together an interdisciplinary, multi-university project involving two ME students and two CE students from Purdue and two EE and two Biology students from UPRM. The Purdue students visited PR at the beginning of the semester to define the project and build the team. During the semester the project was performed via distance collaboration, and then the Purdue students joined the team for a final design presentation to the community leaders, who are now raising funds to implement the design.[6]

Another project I advised is a solar oven for use in Tanzania, an ME student project in collaboration with a non-profit, Solar Circle, dedicated to improving life in Tanzania. The oven design requirements involved manufacturing in Tanzania with local materials and processes for $30U.S. The two students who went to Tanzania, Matt (chief engineer on the project, chosen by the student team) and Jeff, were quite special. These two organized a team for a summer offering of our senior capstone design course. There had been a previous semester of work on the project, and I put out the word I was looking for volunteers to continue the project. Their prototype underwent a very successful demonstration at the end of the summer term, in fact, so successful that an invitation to present their design at the Solar Oven Technology Workshop in Morogoro, Tanzania in September 2007 was forthcoming. The organizers included a colleague, Professor Brian Thompson from Michigan State, as well as Judy Martin from Solar Circle. Travel support came from Shell Co. and a generous alumnus of the ME School, Ken Decker who had been involved in a similar project in Liberia. Matt started a job in August, but he took time off without pay to participate in the workshop. The

---

[6] O'Neill-Carrillo et al., "Mentoring Interdisciplinary Service Learning Projects," 2007.

Purdue Solar Oven was tested in Tanzania along with others from the workshop, with some features incorporated into an ongoing design. Global service learning projects definitely have a great impact on those who do them, and, hopefully, are of sustainable benefit to the "customers" as well. It is essentially impossible to work with the developing world and not learn from experience that many technologies that are ubiquitous in the U.S. are not appropriate elsewhere.

CONCLUSION

What I wish for each of our Purdue students is the opportunity to share the breadth of experiences I have been able to have during my own career trajectory. As challenging is that is, we have no choice. Working with very interesting and very different people from around the world is exhilarating. But more importantly, peace and prosperity for future generations will depend on shared solutions to global grand challenges, and the interconnectedness that comes from international collaboration.

In my judgment, engineers and the firms/organizations they represent will either be global leaders or regional followers. My personal mission is to ensure that Purdue engineering educates global leaders and makes a global impact through research contributions and engagement with engineers and communities from around the world. Since a substantial part of the future of the human race is in the hands of the engineers being educated today around the globe, having students from around the world working together on co-located or distance teams is the best way to educate global engineers and global citizens. A well-trained global engineer has experience making real collaborative decisions on real design projects in which diverse cross-cultural perspectives had to be understood and integrated into the process and product for them to be successful.

REFERENCES

Chiu, G., E. A. Groll, and E.D. Hirleman. "The Purdue Global Engineering Alliance for Research and Education (GEARE) Program." Paper presented at the 2007 ABET Annual Meeting: *The Global Workforce: The Future of Technological Education.* 102

Groll, E. A. and E. D. Hirleman. "Undergraduate GEARE Program: Purdue University's School of ME Contribution to Educating Globally Sensitive and Competent Engineers." Paper presented at the 6th ASEE Global Colloquium on Engineering Education, Istanbul, Turkey, 2007. 102

Hirleman, E. D. "Chemi-Ionization in Homogeneous Methane Oxidation." M.S. Thesis, Purdue University, West Lafayette, IN, 1974. 95

Hirleman, E. D. and S. L. K. Wittig. "Microwave Measurements of Non-Equilibrium Electron Densities in Reacting Gas Mixtures." In *Modern Developments in Shock Tube Research*, edited by G. Kamimoto, 734. Kyoto, Japan: Shock Tube Research Society, 1975. 95

Hirleman, E. D. and C. F. Bohren. "Optical Particle Sizing: an Introduction by the Feature Editors." *Applied Optics* 30 (1991): 4685–4687. DOI: 10.1364/AO.30.004685 98

Hirleman, E. D., E. Groll, and D. Atkinson. "The Three Axes of Engineering Education." Paper presented at the 2007 International Conference on Engineering Education (ICEE). 93

O'Neill-Carrillo, E., L. Seijo, F. Maldonado, E. D. Hirleman, E. Marti, and A. Rivera. "Mentoring Interdisciplinary Service Learning Projects." Paper presented at the 2007 Frontiers in Education Conference. DOI: 10.1109/FIE.2007.4418137 106

Zallen, D. M., E. D. Hirleman, and S. L. K. Wittig. "Determination of Contact-Region Gradients and Their Influence on Shock Tube Properties." Paper presented at the 24th Annual American Physical Society (APS) Meeting, Fluids Division, Boulder, CO, November, 1972. New York: APS, 1972. 94

CHAPTER 5

In Search of Something More: My Path Towards International Service-Learning in Engineering Education

Margaret F. Pinnell

SUMMARY

My personal and professional travels led me toward global education, in particular the University of Dayton (UD) program Engineers in Technical Humanitarian Opportunities for Service-Learning (ETHOS). I do not believe I ever intended to become part of an international engineering education program. In fact, I think the international piece was just a bonus. What really drew me to the ETHOS program was what initially draws our students to the program – a desire to "help" people. In order for the reader to understand how I ended up being involved in global education through the ETHOS program, I must first explain where I came from, my rather unusual career path and the Marianist Spirit that is UD. The reason for this is that my upbringing makes my later involvement in an international program somewhat ironic. My very early experiences as a practicing engineer guided me towards the path of engineering education in the first place and, in particular, my approach towards engineering education. My somewhat unusual career path had a huge impact on what I value as an engineer and educator and my willingness to take career risks as an untenured professor at UD. These career risks came in the form of breaking many of the unwritten (and sometimes even clearly stated) rules associated with the pursuit of an academic career. Most importantly, however, the unique culture at UD provided me with the passion, opportunity and support to not only grow the ETHOS program but to also grow with the ETHOS program.

INTERNATIONAL ROOTS - HOMETOWN PULL

I was born and raised in Youngstown, Ohio. Like so many people from Youngstown, all four of my great grandparents came to the United States from a foreign country as young adults to work in the steel mills. Because of its roots, Youngstown was and continues to be, richly ethnic, benefitting from its Italian, Slovak, Polish, Swedish, and Irish roots. However, once families settled in Youngstown they typically did not leave or even travel very far. It was quite common for large extended families, like mine, to live within a few miles of each other. There was no real reason for anybody to leave Youngstown. Families were close by and the community offered just about whatever a person might need. As a result, most people from Youngstown stayed close to home, attended the local university or trade school then worked and raised their families there. So, even though I enjoyed a variety of authentic international foods and traditions growing up, the idea of international travel was not on my radar.

AN UNUSUAL PATH TO AN ACADEMIC CAREER

As mentioned above, I had a slightly unusual career path that included what many may consider, "risky career choices." These career risks included receiving all three degrees from the same university and then teaching at that university, taking time off to be a stay at home mom, and becoming very much involved in service-learning as an untenured professor. Since I never would have gotten involved in the ETHOS program had I not taken these career risks, I feel the need to go into a bit more detail.

CAREER RISK 1 - ALL THREE DEGREES FROM THE SAME UNIVERSITY

I am the product of my own, Catholic, Marianist university. I teach at the same university, the University of Dayton (UD), where I received all three of my degrees. Getting three degrees from one university or, even more unusual, pursuing a career at the university from which you received all your degrees is something that is certainly not recommended and is, in fact, prohibited at some universities. In my own defense, however, I never really planned on earning all three degrees from the same university. It was never really my intention, in fact, to pursue an academic career.

I ended up being an engineer by default. For as long as I can remember I wanted to be a large animal veterinarian. This career path stemmed largely from my love of animals, in general, and my love of horses, in particular. Unfortunately, however, my dream of becoming a horse vet was crushed when I passed out three times in three days while volunteering at a vet clinic when I was in the ninth grade. Realizing that I did not have the stomach to be a vet, I was forced to consider alternative career paths. At that time, I was sure that I did not want to be a teacher like my Mom or a carpenter like my Dad. Furthermore, having grown up relatively "poor," I wanted a career that would provide me with some financial security. Like many students who end up in engineering, I was good in math and science. Because of this, my Mom suggested that I go into engineering. Although I really had no idea what engineering was all about, I decided to heed my Mom's advice because her advice was always good.

The first degree I received from UD, a Bachelors in Mechanical Engineering (BME), was the product of choice. My guidance counselor informed me of UD's great reputation for undergraduate engineering education. Even though UD was a little far from home (250 miles), it was far more affordable than some of the other universities I considered. UD's affiliation with the Catholic Church also appealed to me. I grew up with a great deal of faith but very little religion. By this, I mean that my family instilled a great love for God and sound foundation of Catholic values, but we rarely went to church. This changed when I was sixteen and able to drive myself to church. I started attending our local Catholic Church on a regular basis. I believed that going to a Catholic university like UD would enable me to learn more about my faith and also enable me to be more actively engaged in the Church.

The biggest selling point of UD, however, was the "feel" of the university community. When I stepped foot onto UD's campus for a visit, I was immediately captivated by its friendly atmosphere and great sense of community. I felt at home. This uniquely warm, welcoming and nurturing atmosphere is something that is very difficult to put into words, but something that so many people who have visited UD have experienced and have also tried to express. Although not helpful to those who have never experienced it, I believe that this atmosphere may best be described as the "Marianist Spirit." The best I can do to define this Marianist Spirit is to list the five characteristics of a Marianist Education which include: (1) educate for formation in faith, (2) provide an integral quality education, (3) educate in the family spirit, (4) educate for service, justice and peace, and (5) educate for adaptation and change.[1] Powerfully charismatic, this Marianist Spirit is what attracted many of my classmates to UD and continues to attract prospective students, including my son, Ronnie, as well as faculty and staff to UD.

My second degree from UD, an M.S. in Materials Engineering, was the product of convenience. As an undergraduate student, I co-oped for five semesters during which I always took evening classes. As a result, I happened to accumulate enough graduate credits as an undergraduate student that getting my master's degree only required three additional months of full-time study. Additionally, staying at UD to finish my master's degree provided me with an opportunity to continue working with an excellent researcher, mentor, and teacher, Dr. Peter Sjoblom from the University of Dayton Research Institute (UDRI).

My third degree, a Ph.D. in Materials Engineering, was the product of "necessity." About two years after I received my Master's degree, my husband, Brad, and I decided to start our family. Even before we got married, we agreed that one of us would stay home with our children while they were young. I really wanted to be the one who got to be the stay at home parent, but at the same time, I did not want to become so out-of-touch with the engineering profession that I could never return. In considering my career options, I reflected on the impact that some of my engineering professors had on me throughout my five years as an undergraduate student. There were many occasions where my professors had gone out of their way to help me succeed in something I pursued. The notion of being able to have this type of impact on the life of a young person appealed to me. As a result, I decided that being a "teacher" like my Mom might not be so bad after all. In fact, I became quite passionate about becoming a college engineering professor. Becoming a college professor, however, required a Ph.D. In my overambitious twenty-something-year-old mind, I reasoned that getting my doctorate while having my family might be a good way to stay in touch with the engineering profession while also providing the credential I needed to become a college professor. After all, my

---

[1] Marianist, *Marianist blogspot*, 2009.

Mom had earned her teaching degree while raising my brother, sister, and me AND working at a factory.

By the time I had quit my job and decided to pursue my doctorate, I was seven months pregnant with my son, Ronnie. My husband, Brad, had an exciting job that he was uniquely qualified for working as a failure analyst for UDRI. Being one of the most selfless people I know, Brad was willing to give up his job and move so I could go to a different university. However, I could not let him make this sacrifice as he loved his job more than anybody I had ever met. Furthermore, my husband's entire family lived close by and my family was "only" 250 miles away. I did not want my children to grow up not knowing their grandparents, aunts, uncles, and cousins. So, while I might say I had to stay in the area, in reality, I wanted to stay in the area. Most importantly, the impending birth of my son and my desire to be a stay-at-home mom required me to find flexibility in a graduate program. The Marianist Spirit of UD includes catering to the individual needs of the students. As such, it was really no surprise to me when the director of the Graduate Materials Engineering Program at UD, Dr. Jim Snide, expressed a willingness to work with my personal situation and goals. Not only did he allow me to do most of my work and research from home, but he also provided me with the tools and equipment that enabled me to do this. With the emotional, moral, practical, and financial support of my husband, I started my degree just six days before my son Ronnie was born, took my qualifying exams two weeks before my daughter Erin was born and defended my dissertation just a week and a half before my youngest daughter, Marie, was born.

After earning my doctorate, I chose to work for UD because I continued to be captivated by its Marianist Spirit and because I had developed an even deeper sense of commitment to the University. UD had treated not only me, but also my husband very well. My post Ph.D. career led me "back" to UD, first as a part time researcher for UDRI, then as an adjunct professor, visiting assistant professor and currently (but hopefully not for too much longer) assistant professor.

CAREER RISK 2 – TAKING TIME OFF TO BE A STAY-AT-HOME MOM

After receiving my Master's degree, I accepted a job at the same defense-related organization where I co-oped as an undergraduate student. My professional duties, however, were significantly different from those I had experienced earlier. As a co-op I did materials-related research and spent a great deal of time working in the laboratory and working under the supervision of an excellent teacher and mentor. As a professional however, I was assigned to program management. Not being able to get into the lab and do research and having to spend a great deal of time on paper work was not appealing to me. Program management seemed very dull to me. More importantly, when I was asked to manage programs developing war-related materials, I also found my job to be in contradiction of my pacifist values and beliefs. How could I work on a program that seemed warlike to me? I have great appreciation and respect for the abundance of quality research, scientific advancement, and innovation that comes from defense-related programs, but I could not separate myself from the fact that the work I was doing was for military applications. "War" work was just not for me. I came to realize that I needed more than a job or even a career – I needed a vocation. I needed to find a way in which I could use my engineering talents and skills to serve God by helping humanity.

Since I was not happy with my job, my husband and I decided that it might be a good time to start our family. So, after working for only two years, I made a major career shift and became a stay-at-home mom. During the ten years that I was a stay-at-home mom, I worked part-time in the evenings towards my Ph.D. and, eventually, as a quarter-time researcher at UDRI. Although many may not see the value of being a stay-at-home mom from a career-development standpoint, it was probably one of the most profound career-development activities I have ever experienced. Being home with my children taught me a great deal about time management, multi-tasking, burning the midnight oil to complete a project, how to find creative solutions to interesting problems, and the importance of communication. Most importantly, it helped to order my priorities of faith, family, and friends, and it helped me to paint a clear picture of the type of engineer I wanted to be – one who made a difference in the world and engaged in activities and research that were in alignment with my core Catholic values, including a commitment to social justice and peace.

CAREER RISK 3 – SERVICE-LEARNING BEFORE TENURE

When I finally started teaching at UD, I was hired on as a part-time adjunct professor. Although the position of adjunct professor does not provide professional security, it also does not have the same requirements and hurdles as a tenure-track position. I felt I had a great deal of freedom to put energy into the aspects of teaching that really interested me and to pursue things I felt had great value. My position was later changed to a part-time visiting assistant professor and later to an assistant professor. As luck would have it, two of the first people I met after becoming a visiting assistant professor at UD were Dr. Dave Herrelko, the director of The New Engineer Program, and Chris Schmidt, a graduate assistant that was hired to help facilitate the newly formed Engineers in Technical Humanitarian Opportunities of Service-Learning (ETHOS) Program.

Dr. Herrelko, introduced me to the concept and practices of service-learning. Some of the main objectives of the New Engineer Program were to provide opportunities for undergraduate engineering students that would help them to develop "soft" skills such as team work, communication, and leadership and to gain a better understanding of the relationship between engineering and society in the context of the Catholic faith in the Marianist tradition. The program offered faculty mentorship, dinners with professionals, special workshops such as interviewing skills, opportunities for students to attend cultural events in the community, scholarship money to take classes not related to engineering, and many other activities. A major component of the New Engineer Program was service. The first-year engineering students completed a large service-learning project as part of an Introduction to Engineering and Design course. They learned about the engineering in bicycles, collected used bikes, and then hosted a Wheels Day for deserving youth in the community. On the Wheels Day, the first-year engineering students worked with the youth to fix up a bike, learn about bike safety, and have some fun through a bike rodeo. I was very impressed as I could see the impact this program had on both the technical and the non-technical learning of the students. Students who participated seemed more motivated to learn and more excited about engineering, and they had an opportunity to practice important non-technical skills such as communication and teamwork. Furthermore, I was delighted that the students were provided with the opportunity to see how they could use their engineering skills to help people and become involved in their community. I could not help but think that the students had been given a special gift – a deeper understanding of what one could do with an engineering degree and an opportunity to view engineering as more than just a career, but instead as a vocation.

The graduate student I met, Chris Schmidt, was running the newly created ETHOS international service-learning immersion program. The program involved preparatory sessions followed by a ten to sixteen week service-learning internship with an NGO or other organization that worked with appropriate technology in a developing country. Once again, I could immediately see the value of this program and wished I had the opportunity to participate in such a program when I was a student.

Reflecting on the value of both service-learning and the ETHOS program, I decided to implement an ETHOS related service-learning project in my materials laboratory course. Everything about service-learning made sense to me. It was a great way to "multi-task," something I learned to value greatly as a stay-at-home mom. Students could learn by doing something useful and not "waste" resources or valuable time. The community partner could benefit from the work done by the students. Instead of conducting a lab experiment on test samples purchased just for that lab exercise so students could break them and write a report only I would read, the students were now testing materials for eco-efficient cook stoves and providing that information to a non-governmental agency that would not only read the report but also make use of the data. Additionally, the students were learning about another culture and about life in the developing world. They were able to use their engineering talents in ways that helped humanity and to see the impact of engineering decisions on the world around them. The service-learning project had the added benefit of providing a few of the students with the opportunity to present their work at a "stovers" conference, which was an excellent experience for them. Service-learning seemed like a win-win situation to me. Although facilitating this project took a great deal of time, I found it to be time well spent as all of the stakeholders (students, "community" partner, and the university) seemed to benefit greatly from this work.

As a result of this project, I became very interested in the pedagogy of service-learning as well as appropriate technology and international development. I felt I had found my calling as well as my niche in engineering and in engineering education – a far cry from the way I felt as a program manager on a war-supporting research program. My interest in service-learning and appropriate technology motivated me to seek out additional ways to become involved in the ETHOS program as well as in service-learning and appropriate technology. Within a short time, I became more and more involved with ETHOS and eventually became its faculty director (in part because no one else really wanted to do it).

If earning my doctorate taught me anything, it was to never take things at face value and to question everything. As a result, neither my "motherly" intuition nor my professional intuition was enough to convince me that service-learning was a sound pedagogy. I felt the need to verify my intuition through research. I started to dig through the vast amount of literature that was written on the topic. In doing this, I learned a great deal about the concept and practice of service-learning, as well as the advantages and disadvantages of this emerging pedagogy, especially in the context of engineering education.

Although I found many definitions of service-learning in the literature, all of these definitions include the integration of "community" based service projects into the curriculum with the dual goal of helping students learn course content through experiential learning and helping a service-organization or community partner meet a particular need. Through my research I found out that in order to achieve these goals, a properly-facilitated service-learning project must include three important components: "course" structure, community service, and structured reflection.[2]

Many of the articles I read identified some of the same key benefits of service-learning that I had observed through my own work. Among these key benefits were helping students to develop technical and non-technical skills; build self confidence; enhance problem solving skills, creativity, and ability to adapt to difficult situations; make connections between classes; develop racial and cultural sensitivity; enhance a commitment to civic responsibility; and increase ethical awareness and awareness of the impact of professional decisions on society and the environment.[3] More importantly, I have seen students grow and transform through participation in service-learning projects. I have also seen students make life and career decisions based on these experiences.[4] I have seen great partnerships and friendships develop between students and representatives from the community being served as well as partnerships between faculty members and students, all as a result of service-learning.[5]

Although my experiences resonated with the many educational advantages of service-learning presented in the literature, my research also provided me with a reality check in that it alerted me

[2] Duffy, "Service Learning in a Variety of Engineering Courses," 2000; Eyler, "What I Never Learned in Class," 2000; Gordon et al., "Rice University Engineers Without Borders," 2006; Husfeld et al., "Lessons Implemented on an International Service Learning Project," 2005; Lucena and Downey, "Engineering Cultures," 1999; McCarthy and Tucker, "Student Attitudes toward Service-Learning," 1999; Morton, "A Smart Start to Service-Learning," 1996; Paterson et al., "International Service Engineering Across Academic Borders," 2006; Riley and Bloomgarden, "Learning and Service in Engineering and Global Development," 2006; Tsang, "Service-Learning Pedagogy for Engineering," 2000.

[3] Bordelon and Phillips, "Service-Learning," 2006; Duffy, "Service Learning in a Variety of Engineering Courses," 2000; Eyler, "What I Never Learned in Class," 2000; Frank et al., "Implementing the Project-Based Learning Approach in an Academic Engineering Course," 2003; Jamieson et al., "EPICS," 2001; Gokhale and O'Dea, "Effectiveness of Community Service in Enhancing Student Learning and Development," 2005; Gordon et al., "Rice University Engineers Without Borders," 2006; Husfeld et al., "Lessons Implemented on an International Service Learning Project," 2005; Jawaharlal et al., "Implementing Service learning in Engineering Curriculum," 2006; Lucena and Downey, "Engineering Cultures," 1999; McCarthy and Tucker, "Student Attitudes Toward Service-Learning," 1999; Morton, "A Smart Start to Service-Learning," 1996; Oakes et al.," EPICS," 2001; Paterson et al., "International Service Engineering Across Academic Borders," 2006; Riley and Bloomgarden, "Learning and Service in Engineering and Global Development," 2006; Tsang, "Service-Learning Pedagogy for Engineering," 2000.

[4] Eger et al., "Student Perspectives of Curriculum Integrated International Technical Immersions," 2006; Kleinhenz et al., "Student Perspectives of Curriculum International Service-learning Internships," 2005; Pinnell et al., "Innovation, Entrepreneurship and International Experience," 2008.

[5] Pinnell et al., "Evolution of a Multi-Disciplinary Service-Learning Project with a Community Partner," 2005; Pinnell and Chuck, "Developing Technical Competency," 2004; Pinnell et al., "International Technical Service Immersions," 2007; Pinnell et al., "Innovation, Entrepreneurship and International Experience," 2008.

to some of the challenges and limitation of service-learning. I did not think much about these challenges before I became involved in service-learning, and I doubt that knowing about them would have deterred me from pursuing service-learning projects. Regardless, they do exist. One challenge presented in the literature, which I have also noticed, speaks to the problem of scale - not only in the quantity of service-learning projects, but also the quality of these projects. Only a relatively small number ($< 25\%$) of students end up participating in service-learning projects, even at universities that proclaim to have service-learning at the core of their institutional structures. Disturbingly, many students who have been involved in service-learning projects do not see the value of this pedagogy because of the way the project was facilitated or the "quality" of the project. One reason for this is the lack of a clear definition of what service-learning is and is not. Great inconsistencies exist in how service-learning projects or courses are developed and implemented. In particular, many community partners do not have a full understanding of the reciprocal nature of service-learning. For example, the community partner in a service-learning project I facilitated in one of my engineering courses failed to provide some of the communication as well as facilities required to complete the project. This had a negative effect on the students' learning as well as the students' perception of the experience. Because a widely-shared understanding of service-learning does not exist, students have a broad range of good and bad experiences with service-learning. Equally important, community partners can also have a broad range of good and bad experiences with service-learning and can sometimes feel exploited for the benefit of the students.

Another challenge with service-learning is the ethical and liability risks that exist with students (not degreed professionals) doing work in the community and/or in other countries. Safety and liability issues also exist when instructors take students off campus to work with tools and in locations that pose potential health or safety hazards. Even if these liability issues did not exist, the personal responsibility I take when sending students out to work in the community can often times be very stressful.[6]

Probably the challenge that I was warned the most about was the negative effect that being involved in service-learning could have on my ability to get tenure. It takes a great deal of time and sometimes money to implement a service-learning project in a course. The time required to implement these projects may take away from the time that an untenured professor should be spending doing research and writing journal articles and grant proposals. Furthermore, implementing a service-learning project can have a negative impact on teaching evaluations. At a workshop I co-organized in 2004 at the University of Colorado at Boulder with some very well-respected professionals who were actively engaged in service-learning, I was warned on numerous occasions that I should get tenure first and then worry about being involved in service-learning.[7] Being somewhat stubborn and idealistic, I did not heed that advice. At the time that I was given this

---

[6]Banzaert et al., "Faculty Views of Service Learning in Mechanical Engineering at MIT," 2005; Bordelon and Phillips, "Service-Learning," 2006; Hinck and Brandell, "The Relationship Between Institutional Support and Campus Acceptance of Service-Learning," 2000; Jawaharlal et al., "Implementing Service Learning in Engineering Curriculum," 2006; McCarthy and Tucker, "Student Attitudes Toward Service-Learning," 1999; Ward, "Addressing Academic Culture," 1998; Zlotkowski, "A Service Learning Approach to Faculty Development," 1998.
[7]Bielefeldt et al., "Creating an EDC Emphasis in Environmental Engineering," 2005.

advice (2004), I had been actively engaged in service-learning through ETHOS and course based service-learning projects for nearly three years. I had become involved in these activities as a visiting assistant professor, when tenure was not a concern. Furthermore, even after I was placed on the "tenure track" in 2003, I did not have a full understanding of what was required to get tenure at UD nor how being involved in service-learning might be a detriment to achieving that goal. I felt surely that, at my beloved Catholic, Marianist institution whose motto is "learn, lead and serve" and where a commitment to social justice runs through heart and soul of the university, being involved in service-learning would be valued and not hinder my ability to get tenure.

Although my mid-term review, which was completed in 2006 by our departmental promotion and tenure committee, did not show overwhelming support for my service-learning research or work, the administrative support from both the Dean, Dr. Joe Saliba, and my department chair, Dr. Kevin Hallinan, for the ETHOS program and many of the service-learning projects I have facilitated has been tremendous. Their support suggests that this work is valued by both the School of Engineering and the university. Furthermore, in 2004, I was awarded the School of Engineering Award of Excellence for Teaching, and in the spring of 2008, I was awarded the Alumni Award in Teaching, largely due to my involvement with the ETHOS program and service-learning.

Despite the challenges and limitations associated with service-learning, I am of the opinion that the advantages greatly outweigh the disadvantages. Regardless of how my involvement has affected me professionally, I know for a fact that, much like my decision to stay at home with children, I would not, for a minute, alter my decision to be involved in ETHOS and in service-learning. I had learned through my early career experiences and my lifestyle choices the importance of making decisions and doing things that are in alignment with my values and priorities, even if that comes with personal and professional risks and sacrifices.

PARADIGM SHIFT IN ENGINEERING EDUCATION

As an engineering student in the 1980s, I heard very little, if anything, about international education, experiential, or service-learning. We were taught engineering through many lecture based courses and a few labs. We were taught to apply formulas to solve engineering problems from the book. We rarely worked on projects and, except for a few lab classes, we rarely were expected to work in teams. Our exams, although very challenging, generally consisted of solving a few problems that were no more than an extension of the problems we did as homework. We did not have to learn how to communicate with people through writing or through verbal communication, and we certainly were not expected to know anything about the world or travel. Most of the experiential learning my peers and I received was through co-op positions. Many of these co-op positions were with automotive related manufacturing companies that were thriving in Dayton at that time or with the research labs at Wright-Patterson Air Force Base. Upon graduation, we were expected to get a job with a company (most likely, a manufacturing, defense, or space industry), agency, or go to graduate school.[8] When reflecting on this, I thought that maybe these observations of undergraduate engineering education were uniquely mine. However, by conducting an e-mail interview of some of my engineering classmates, I found they had similar memories of their undergraduate engineering education.[9] I then began to wonder if these experiences were unique to students at UD. A review of the literature, however, indicated that the lecture-based pedagogy I had experienced as a student was typical for engineering education in the 1980s.[10]

When I became an adjunct professor in the late 1990s, it did not take me long to realize that engineering education had changed since I was a student. I observed that many of the incoming engineering students at UD were from middle to upper middle class suburban "white collar" families as opposed to the rural and "blue collar" families of many of my classmates in the 1980s. Many students had grown up playing video games, many had attended academically challenging high schools that offered high-level math and science classes, and most were involved in numerous organized activities while in high school. Many, however, had very little, if any, hands-on experience in fixing things or in "tinkering."

Furthermore, the face of engineering had changed quite a bit. Computers, the information super highway, and e-mail had transformed how the students, faculty, and professionals accessed data, solved problems, and communicated.[11] Our society was experiencing an ever-expanding global economy and jobs in the engineering profession were changing from ones based in manufacturing and construction to ones that were largely service based.[12] Furthermore, the incoming students I met expressed a desire to do more than just make money in their chosen career—they wanted to make a difference in the world. Therefore, students not only expected to be taught differently but they had to be taught differently in order to get a job they would view as fulfilling after graduation.[13] Lucky me! I felt that the changing needs of the students and the new Accreditation Bureau of Engineering and Technology (ABET) requirements were right in line with my priorities and the outlook on engineering that I had developed during my early career experience and my time as a stay-at-home mom.[14] To me, this seemed like a very exciting time to be part of the engineering academic community! As a result, I continued my research on the changing face of engineering education, and for those who share a similar passion, I include a rather extensive bibliography at the end of this chapter.

---

[8]Hodge and Taylor, "Factors for Change in Mechanical Engineering Education," 1997; Kenyon, "The Coming Revolution in Engineering and Engineering Technology," 1993.

[9]E-mail Interview, 2008.

[10]Hodge and Taylor, "Factors for Change in Mechanical Engineering Education," 1997; Kenyon, "The Coming Revolution in Engineering and Engineering Technology," 1993; Kolar and Sabatini, "Changing from a Lecture Based Format to a Team/Learning/Project Driven Format," 1997.

[11]Boylan, "Characteristics of First-Time Freshmen in 4-Year Institutions Intending to Major in Engineering and Computer Science," 2003.

[12]Wei, "Engineering Education for a Post Industrial World," 2005.

[13]Katechi et al., "A New Framework for Academic Reform in Engineering Education," 2004; Mina et al., "Learning to Think Critically to Solve Engineering Problems," 2003; Stromquist, "Internationalization as a Response to Globalization," 2007.

[14]Accreditation Board for Engineering and Technology, *ABET Engineering Criteria 2000*, 2000.

One observation I noted, but found difficult to verify through research, was that the career aspirations of students had changed. When I was a student, I do not remember any of my peers specifically seeking a job that would make a difference in the world. Instead, most of my peers, including myself, looked for a stable job at a reputable company that offered good pay and benefits. At that time, the manufacturing sector was fairly strong in Ohio, so many of my peers worked in manufacturing related jobs. With few exceptions, most have changed jobs several times since completing their degrees. In contrast, many of the students I have advised over the past five years have looked for jobs at companies working in areas that help humanity, such as medical device companies or those related to sustainability, or they have taken voluntary service positions for several years after graduation. The only evidence I can find that supports the desire of engineering students to "do good" is the popularity of many service-based engineering programs such as Engineers Without Borders (EWB) and Engineering Projects In Community Service (EPICS). Both of these, as well as other similar programs, have experienced explosive growth in the past seven to ten years and have attracted women in ratios that do not match the demographics of the engineering student body.[15]

GROWING WITH THE ETHOS PROGRAM:

Probably the one factor that has contributed the most to the success of the ETHOS program is that it was founded by students for students. In a sense, it is a living example of "appropriate technology" as this program was not imposed on the students, but instead had very organic roots. The ETHOS program was developed by an interdisciplinary group of undergraduate engineering students in the spring of 2001, as part of a capstone engineering design course. The previous summer (2000), one of the founding students, Christine Vehar, had participated in an international service immersion placement in India, sponsored by the University's Campus Ministry program. Her work involved tutoring and working with children. Although Christine felt this immersion was a wonderful and life changing experience, she was frustrated that she was not able to make use of her engineering skills while there, especially since she could see how many of the humanitarian issues that existed at her placement site could easily be addressed with rather simple engineering interventions. Christine approached, the Mechanical Engineering department chair, Dr. Kevin Hallinan, with these concerns. Dr. Hallinan challenged her to create a service-learning immersion program specifically for engineers. To facilitate development of the program, Dr. Hallinan provided financial sponsorship as well as mentorship and guidance. Generally, capstone design projects address industrial needs and are sponsored by local companies. Although this alternative would be considered unique at many other universities, at UD, it really just represented another way in which a professor nurtured the interests and intellect of a student. It simply echoed the Marianist Spirit of UD. The founding students were able to use the engineering design process to research and develop the program, establish its name and acronym, and make initial placement contacts with partner organizations.

[15] Barrington and Duffy, "Attracting Underrepresented Groups to Engineering with Service-Learning," 2007; Bielefeldt et al., "Creating an EDC Emphasis in Environmental Engineering," 2005; Engineers Without Borders, 2007; Gordon et al., "Rice University Engineers Without Borders," 2006; Jamieson et al., "EPICS," 2001; Oakes et al., "EPICS," 2001; Pinnell et al., "International Technical Service Immersions," 2007; Purdue University, *EPICS*, 2007.

Part of the research the students conducted included identifying and studying similar programs already in existence (at that time EWB was just being formed). In developing ETHOS, the students worked with UD's Campus Ministry to adapt an existing immersion preparatory short course to meet the needs of the engineering students who would be participating in this program. The founding students spent a great deal of time consulting with Brother Phil Aaron who ran the immersion program through Campus Ministry. Brother Aaron cautioned the students about "serving" versus "learning" and the potential for service-projects to have negative effects (economical, self-sufficiency, cultural) on the community being served. The students also used lessons learned from failed service projects at other universities to create a program that would not exploit people or communities in developing countries or suggest inappropriate or unsustainable solutions to the humanitarian issues the students would try to address. The president of UD at that time, Brother Ray Fitz, also shared a great wealth of experience with the students and provided support and guidance in the development of the ETHOS program.

The mission statement developed by the students for this program reads, "ETHOS was founded on the belief that engineers are more apt and capable to serve the world more appropriately when they have experienced opportunities that increase their understanding of technology's global linkage with values, culture, society, politics, and economy." Much of the excellent research and work that the founding students completed serves as the foundation for the ETHOS program today.

When I first became involved in ETHOS in the fall of 2001, the University was preparing to send its first set of five students on their technical service-learning immersions. At that time, no academic credit was awarded for this experience, few students were aware that the program existed, only a few students expressed interest in participating, and most of the travel arrangements and preparatory sessions were coordinated by the graduate assistant, Chris Schmidt. The graduate assistant was partially funded through the Mechanical Engineering Department but also had a research assistantship with a faculty member. Funding for the student travel was provided by various internal funding sources including the School of Engineering, Campus Ministry, Chaminade Scholar Program, and the New Engineer Program, and through personal and group fund raising efforts. As the program grew, I willingly, and with great enthusiasm, became more and more involved.

Within a short time, I became the unofficial director of the program. I worked with the various graduate assistants (they change every one to two years) to develop an elective course that formalized the preparation process, created a clear definition of the student deliverables, and provided participating students with academic credit. Additionally, the graduate assistant and I formalized the paperwork; created a handbook for the course, brochures, and a website; and we hosted seminars to advertise the program. We worked with the newly appointed engineering Dean, Dr. Joe Saliba, who committed funding and other resources to the program. We also wrote proposals to seek external funding for the program; wrote papers, and conducted research on service-learning, engineering education, appropriate technology and international education; participated in workshops and conferences; and made other efforts to collaborate and communicate with other professionals involved in similar programs and work. Within a few short years, the graduate assistant (Charlie Schreier) and

I developed an on-campus ETHOS club to provide awareness about global engineering issues and appropriate technology and to provide local service-learning opportunities as well as international spring break-outs for engineering students. As the program became more popular, we developed an application and selection process for students interested in participating in an ETHOS immersion. As the Center for International Programs (CIP) at UD started to grow and become much more organized, we started to collaborate with their talented and committed new director, Amy Anderson, and the rest of the CIP staff to take advantage of the many resources they offered, such as safety information, passport fairs, on-line student registration, Study Abroad Fair where we were able to host a booth, and much more. We also worked with our development office to get funding for the program through private donors. Although much of the coordination work is still done by the graduate assistant, my role as faculty director is to provide continuity and growth to the program, seek funding, help coordinate the preparatory classes, provide the students with grades, seek new partnerships, and serve as advocate for the student participants. A third and very important member of the administrative team, the service-learning coordinator for the School of Engineering and the former consultant to the founding students, Brother Phil Aaron, was added in 2006.

Although ETHOS strives to offer a wide range of activities and experience, the part of the program that requires the biggest investment of my time, as well as the time of the graduate assistant and service-learning coordinator, is the international technical service-learning immersions. However, these immersions also have the biggest impact on the attitudes and perceptions of students. Since its inception in 2001, ETHOS has sent more than 120 engineering students as well as several students from other majors (Geology, Communication, Business, Computer Science) and even one student from a neighboring university to developing countries to work with development organizations and communities. Some of the host countries include Nicaragua, Honduras, Brazil, Bangladesh, Guatemala, Mexico, Bolivia, Cameroon, Togo, India, and Peru. The ETHOS international technical immersions are typically ten to sixteen weeks in length, during which time the students work with collaborating organizations and communities to assist in finding appropriate, sustainable, and effective solutions to technical challenges through guided research, development, and projects. Students generally travel in small teams of two to three students each. Since the students are required to do technical research prior to travelling, are placed with organizations that have an onsite technical director, and are involved in ongoing research and development work, the students are able to contribute individually, or as a team, to an existing or new project in the short amount of time that they are at their placement. In many instances, the students have even continued to work on these projects upon their return to UD then participate in another immersion the following summer at the same placement site to continue their research in a particular area.

Success of the ETHOS program relies heavily on well-established international partners. Because we send students (for the most part) to the same placement sites year after year, there is a clear understanding between the ETHOS program and the international partner regarding the responsibilities of the various stake holders. Not only do the international partners provide technical guidance and mentorship, but they are also responsible for taking care of some of the logistics such as identifying suitable host families and setting up the payment structure for the room and board. Most of the well-established international partners are located in rural villages and small towns. Students typically describe the people in these communities as "poor." During the preparatory training sessions, we stress to the students the importance of establishing relationships in the community and learning with the community about a specific technology. Therefore, interaction with the host families and the communities on a daily basis is integral to a successful immersion experience. Sending students in small teams as opposed to large teams encourages the students to completely immerse themselves in the community. As a result, the students end up learning a great deal about the culture, including the local resources and skills of the community, and they are forced to become proficient in the language spoken there. Most student participants tell us through their reflections and through discussion that their host families and communities were the best part of their trip. We have had students give ballet lessons to the children of a community, help coach soccer, read and play games with the children, and even be "knighted." With the help of the host organization and community, students are able to use their engineering knowledge to address real world problems while learning about another culture and gaining a better understanding of the interface between technology and global society.

Currently, students earn academic credit towards their engineering degrees by preparing for, participating in, and documenting their ETHOS experience. The students participate in a semester-long preparatory course that includes individualized language preparation facilitated through our language department, travel safety and health, practical travel tips such as packing, cultural sensitivity training, an introduction to appropriate technology, and guided research on the work they will be doing at their placement. In the summer, the students participate in their technical service-learning immersion in a developing country. Upon their return, the students are required to write a reflection report and a technical report that summarizes their work, give at least two presentations, attend a dinner with invited faculty and staff, and they also attend a reflection retreat.

Students are encouraged to continue their research when they return to campus and also bring back projects that can be implemented in the classroom. As a result, some of our participants have implemented ETHOS-related research and problems into other classes, such as their honors thesis, experimentation lab, heat transfer, materials, and capstone design experience. One vision that I had for the ETHOS program early on was to have students from other majors become involved in ETHOS-related projects. I also had the vision of having our student participants be able to work on a single ETHOS project throughout their academic career. This vision became a reality when one of our ETHOS participants, Lori Hanna, decided to integrate her immersion experience into her Honors Thesis work and her senior capstone design project. As a result of the work that Lori and her teammates did during their capstone design course, they decided to partner with two business students to enter a business plan competition. The students won this business plan competition and are using their winnings to help finance the start up of a microbusiness, Salud del Sol, that will build and distribute solar autoclaves in Nicaragua and beyond. As a result of this project, the School of Business at UD has committed to sending up to two students each year on an ETHOS immersion with an engineering student.

Personally, I have really enjoyed being involved in the development and growth of the ETHOS program and being able to discuss the wonderful accomplishments of the student participants and graduate assistants with colleagues from UD and elsewhere. Every ETHOS graduate assistant (Chris Schmidt, Bill Eger, Charlie Schreier, and Mike Vehar) has used his own unique gifts and talents to enhance the quality and organization of the ETHOS program. The program has been well received by the School of Engineering, UD, and our international partners. As such, working with this program has been a very positive and rewarding experience with few if any real struggles. I have enjoyed watching the engineering students at UD learn and grow through the international technical service-learning immersions, student organization activities, collaborative research, and hands-on classroom projects that support the development of appropriate and sustainable technologies for the developing world.

INTERNATIONAL EXPERIENCE THROUGH THE ETHOS PROGRAM

Although the aspect that attracted me first to the ETHOS program was the service-learning, I quickly realized the importance of the international experience that is provided through this program. As I mentioned earlier, the engineering profession has changed a great deal since I was a student in the late 1980s. The globalization of the economy makes international experience a necessity for graduating engineers, yet it is not always easy for engineering students, with their packed schedules, to gain such international experience.[16] The internships offered through the ETHOS program represent one approach for engineering students to gain international experience. I have watched students who have participated in this program develop foreign language skills, gain an increased sense of cultural awareness and global civic responsibility, and have a better understanding of the role of engineers in society and in the world. Furthermore, I have seen them become far more confident in their abilities in general, but specifically, in their ability to travel and to live without many of the creature comforts that they had grown accustomed to.

Despite the many educational benefits offered through the ETHOS program, what appears to attract most students is their desire to use their engineering skills to help people and "save the world." Although I believe most students embark on their ETHOS immersion thinking that they will be doing their part to save the world, in the end, it appears that the world is saving the students by providing them with the opportunity to take a risk, step outside their comfort zone, experience a new culture, and see the world and their chosen profession with a new, more mature, set of eyes. It is amazing to me to watch these students transform from somewhat naive, suburban college kids with idealistic values into informed adults who have a greater appreciation of the world around them and an ability to question our traditional way of doing things. Additionally, the ETHOS experience provides the students with a new notion of what it means to be successful. This new definition

[16] Katechi et al., "A New Framework for Academic Reform in Engineering Education," 2004; Klabr and Ratti, "Increasing Engineering Student Participation in Study Abroad," 2000; Stromquist, "Internationalization as a Response to Globalization," 2007; Wei, "Engineering Education for a Post Industrial World," 2005.

no longer has to do with how much money they will make or their position in a company, but, instead, how they might be able to make a difference in our world. Additionally, participating in ETHOS provides many of the students with the opportunity to view engineering as more than just a career—as a vocation.

HOW ETHOS INFLUENCED ME

Going from part-time engineer to full-time mom and then to full-time professor was a difficult transition for me. To be completely honest, and at the risk of being politically incorrect, I found staying home with my children, volunteering in their schools and in the church, and working a few hours from home doing research to be stimulating, challenging, and very personally rewarding. Simply put - I loved it. Prior to having children, however, I dreamed of a career in academia. As mentioned above, this dream stemmed from experiences that I had with professors while I was a student at UD. It seemed that no matter what "out of the box" idea I came up with there was always a professor willing to support that idea and to help me see that idea through. I wanted to be THAT kind of professor. The one who was able to provide the fuel a student needed to turn a spark into a flame - to turn a dream into a reality. As so many professors had done for me, I wanted to be the one to give a student a chance to succeed at doing something they had a passion for, even if that took them on a slightly unusual path. As my youngest daughter, Marie, was preparing to enter the first grade, I was presented with the opportunity to teach materials at UD. This felt like a dream come true for me as I loved the subject and loved UD. The offer was also timely since my children were all in school and I had a little more time to work.

My previous experience in engineering had provided me with a solid notion that I had to pursue areas of research and teaching that were in alignment with my personal beliefs and interests and for which I had a passion for. When I learned about the ETHOS program, it did not take long for me to realize that it was very much in alignment with my personal beliefs and interests. Furthermore, addressing issues of social justice, one of the key aspects of ETHOS, was something for which I had a deep passion. By being in the right place at the right time, I found myself with the opportunity to help grow the program, watch a student's dream of such a program become a reality, and witness how participating in such a program could provide students with life-altering transformative experiences. The ETHOS program gave me the opportunity to use my professional skills to help humanity through the students and to help the students through humanity. ETHOS gave me a great gift; it made my job as a college professor my vocation. ETHOS is the main reason why I love my job!

The future of the ETHOS program looks very bright. Students are interested and engaged and are increasingly able to reach out to the larger student population to share their experiences and to bring others into their experiences. The administration has exhibited growing respect and admiration for the program and continues to support it financially. Because our acting provost and current Dean, Dr. Joe Saliba, has seen the impact that the ETHOS program has had on the participating students and community partners, views the ETHOS program as being in direct alignment with the mission of UD, and understands how ETHOS has provided distinction to both UD and the School of Engineering, he has the vision of expanding the program such that every engineering student is somehow touched by the ETHOS experience. This expansion may include more classroom projects, local and domestic placements, as well as more international placements. Currently, the service-learning coordinator, acting Dean, Dr. Malcolm Daniels, and I are in the process of developing a five year strategic plan for this program.

The stories shared by the ETHOS students through blogs, family letters, TV interviews, local newspapers, campus publications, and ETHOS videos on *youtube* have inspired alumni, friends, family members, and local community members. This has fueled tremendous support for the program through financial contributions, additional collaborative opportunities and, most importantly, prayers. The administrative team continues to work well together and has a shared vision for the program that echoes that of Dr. Saliba's; it is in alignment with the principles of the founding students, and not only upholds, but further enhances the Marianist Spirit of UD.

Although I could never verify this, I believe that it would be very difficult to duplicate the ETHOS program at any other university. I attribute the creation and continued development of this program to the Marianist Spirit of UD. After all, it was the Marianist Spirit, the warm and welcoming campus community, that grabbed my soul when I came for a visit over 25 years ago and still, has not let go. It was the Marianist Spirit of UD, with its nurturing professors, mentors and bosses that encouraged not only me, but the students who founded ETHOS, to pursue our passions, even if that meant taking a "slightly" different path. It is the Marianist Spirit, with its commitment to social justice, that encourages and supports service-learning in general and the ETHOS program in particular. Finally, it is the Marianist Spirit that provides a transformative education for students, creating a network of alumni, family and friends that provide the financial and moral support to the students and to programs such as ETHOS.

I became involved with the ETHOS program because I thought the work the students were doing and the experiences they gained from the program would contribute greatly to the world. But let's face it – I have gained much more FROM the ETHOS program than I have contributed. Being involved with ETHOS has provided me with a way to use my engineering and professorial skills to serve God – at last, I have found my vocation!

Postscript – I received tenure in the winter of 2010.

REFERENCES

Accreditation Board for Engineering and Technology. *ABET Engineering Criteria 2000*. http://www.ele.uri.edu/faculty/daly/criteria.2000.html. Accessed June 7, 2008. 118

Banzaert, A. et al. "Faculty Views of Service Learning in Mechanical Engineering at MIT." Paper presented at the 2005 ASEE Annual Conference and Exposition, Portland, OR, United States, 2005. 116

Barrington, L. and J. Duffy. "Attracting Underrepresented Groups to Engineering with Service-Learning." Paper presented at the 2007 ASEE Annual Conference and Exposition, Honolulu, HI, United States, 2007. 119

Bielefeldt, A. et al. "Creating an Engineering for Developing Communities (EDC) Emphasis in Environmental Engineering." Paper presented at the 2005 ASEE Annual Conference and Exposition, Portland, OR, United States, 2005. 116, 119

Bordelon, T. D. and I. Phillips. "Service-Learning: What Students Have to Say." *Active Learning in Higher Education* 7 no. 2 (2006): 143–153. DOI: 10.1177/1469787406064750 115, 116

Boylan, M. *Characteristics of First-Time Freshmen in 4-Year Institutions Intending to Major in Engineering and Computer Science: An Analysis of Survey Data Collected by the Higher Education Research Institute (HERI)* http://www.heri.ucla.edu/ 2003. Accessed on June 1, 2008. 118

Duffy, J. "Service Learning in a Variety of Engineering Courses." In Projects That Matter: Concepts and Models for Service Learning in Engineering, edited by E. Tsang, 75–79. Washington, DC: AAHE, 2000. 115

Eger, C., C. Schreier, and M. Pinnell. "Student Perspectives of Curriculum Integrated International Technical Immersions." Paper presented at the 2006 ASEE Annual Conference and Exposition, Chicago, IL, United States, 2006. DOI: 10.1109/FIE.2005.1612148 115

E-mail Interview, Results of e-mail interviews of seven of my peers who also attended University of Dayton in the 1980s. May 2008. 118

Engineers Without Borders. http://www.ewb-international.org. Accessed August 10, 2007. 119

Eyler, J. "What I Never Learned in Class: Lessons from Community Based Learning." In Projects That Matter: Concepts and Models for Service Learning in Engineering, edited by E. Tsang, 13–26. Washington, DC: AAHE, 2000. 115

Frank, M., I. Lavy, and D. Elata. "Implementing the Project-Based Learning Approach in an Academic Engineering Course." International Journal of Technology and Design Education, 13 (2003): 273–288. DOI: 10.1023/A:1026192113732 115

Gokhale, S. and M. O'Dea. "Effectiveness of Community Service in Enhancing Student Learning and Development." Paper presented at the 2005 ASEE Annual Conference and Exposition, Portland, OR, United States, 2005. 115

Gordon, R., A. Gordon, and P. Bedient. "Rice University Engineers Without Borders: An Exercise in International Service Learning." Paper presented at the 2006 ASEE Annual Conference and Exposition, Chicago, IL, United States, 2006. 115, 119

Hinck, S. and M. Brandell. "The Relationship Between Institutional Support and Campus Acceptance of Academic Service Learning." American Behavioral Scientist, 43, no. 5 (2000): 868–881. DOI: 10.1177/00027640021955522 116

Hodge, B.R. and R.P. Taylor. "Factors for Change in Mechanical Engineering Education." Paper presented at the 1997 ASEE Annual Conference and Exposition, Milwaukee, WI, United States, 1997. 118

Husfeld, R., C. Polito, and E. Gingerich. "Lessons Implemented on an International Service Learning Project." Paper presented at the 2005 ASEE Annual Conference and Exposition, Portland, OR, United States, 2005. 115

Jamieson, L.H., W.C. Oakes, and E.J. Coyle. "EPICS: Documenting Service-Learning to Meet EC 2000." Paper Presented at the 2001 ASEE/IEEE Frontiers in Education Conference, Reno, NV, United States, 2001. DOI: 10.1109/FIE.2001.963865 115, 119

Jawaharlal, M. et al. "Implementing Service Learning in Engineering Curriculum." Paper Presented at the 2006 ASEE Annual Conference and Exposition, Chicago, IL, United States, 2006. 115, 116

Johnson, E., S. DeMaris, and D. Tougaw. "Providing an Integrated International Experience for Undergraduate Engineering Students at a Small Institution." Paper Presented at the 2006 ASEE Annual Conference and Exposition, Chicago, IL, United States, 2006.

Katechi, L., K. Banks, H. Diefus-Dux, D. Follman, J. Gaunt, and K. Haghighi. "A New Framework for Academic Reform in Engineering Education." Paper presented at the 2004 ASEE Annual Conference and Exposition, Salt Lake City, UT, United States, 2004. 118, 123

Kenyon, R. A. "The Coming Revolution in Engineering and Engineering Technology: A New Paradigm for the 21st Century." Education 113 (1993): 361–371. 118

Klabr, S. C. and U. Ratti. "Increasing Engineering Student Participation in Study Abroad: A Study of U. S. and European Programs." Journal of Studies in International Education Spring (2000): 79–102. DOI: 10.1177/102831530000400106 123

Kleinhenz, P., M. Pinnell, G. Mertz, and C. Eger. "Student Perspectives of Curriculum Integrated International Service-Learning Internships." Paper Presented at the 35th ASEE/IEEE Frontiers in Education Conference, Indianapolis, IN, United States, 2005. DOI: 10.1109/FIE.2005.1612148 115

Kolar, R. and D. Sabatini. "Changing from a Lecture Based Format to a Team/Learning/Project Driven Format: Lessons Learned." Paper presented at the 1997 ASEE Annual Conference and Exposition, Milwaukee, WI, United States, 1997. 118

Leanord, M. et al, "Planning for Curriculum Renewal and Accreditation under ABET Engineering Criteria, 2000." Paper Presented at the 1998 ASEE Annual Conference and Exposition, Seattle, WA, United States, 1998.

Loftus, M. "Cream of the Crop." *ASEE Prism* Summer (2007): 28–33.

Lucena, J. and G.L. Downey. "Engineering Cultures: Better Problem Solving Through Human and Global Perspectives." Paper presented at the 1999 ASEE Annual Conference and Exposition, Charlotte, NC, United States, 1999. 115

Marianist. Marianist blogspot http://usamarianist.blogspot.com. Accessed February 4, 2009. 111

Martin, P. T. and J. Coles. "How to Institutionalize Service-Learning into the Curriculum of an Engineering Department: Designing a Workable Plan." In *Projects That Matter: Concepts and Models for Service Learning in Engineering*, edited by E. Tsang, 41–51. Washington, DC: AAHE, 2000.

Mason, D. E. et al. "Innovation in a Large-Scale Study-Abroad Program in Engineering." Paper presented at the 2004 ASEE Annual Conference and Exposition, Salt Lake City, UT, United States, 2004.

McCarthy, A. and M. Tucker. "Student Attitudes toward Service-Learning: Implications for Implementation." *Journal of Management Education* 23, no. 5 (1999): 554–573. DOI: 10.1177/105256299902300511 115, 116

Mina, M. et al. "Learning To Think Critically to Solve Engineering Problems: Revisiting John Dewey's Ideas for Evaluating the Engineering Education." Paper presented at the 2003 ASEE Annual Conference and Exposition, Nashville, TN, United States, 2003. 118

Morton, K. "A Smart Start to Service-Learning." *Journal of Business Ethics*, 15 (1996): 21–32. 115

Oakes, W., L. Jamieson, and E. Coyle. "EPICS: Meeting EC 2000 Through Service-Learning." Paper presented at the 2001 ASEE Annual Conference and Exposition, Albuquerque, NM, United States, 2001. 115, 119

Paterson, K. et al., "International Service Engineering Across Academic Borders." Paper presented at the 2006 ASEE Annual Conference and Exposition, Chicago, IL, United States, 2006. 115

Pérez-Foguet, A., S. Oliete-Josa, and A. Saz-Carranza. "Development Education and Engineering: A Framework for Incorporating Reality of Developing Countries into Engineering Studies." International Journal of Sustainability in Higher Education, 6 no. 3 (2005): 278–303. DOI: 10.1108/14676370510607241

Pinnell, M. F. and L. Chuck. " Developing Technical Competency and Enhancing the Soft Skills of Undergraduate Mechanical Engineering Students through Service-Learning." Paper presented at the 2004 ASEE Annual Conference and Exposition, Salt Lake City, UT, United States, 2004. 115

Pinnell, M. F. et al. "International Technical Service Immersions: Model for Developing Global Scientists and Engineers in Small to Mid-Size Universities." Paper presented at the 2007 ASEE Annual Conference and Exposition, Honolulu, HI, United States, 2007. 115, 119

Pinnell, M., C. Daprano, and G. Williamson. "Evolution of a Multi-Disciplinary Service-Learning Project with a Community Partner." Paper presented at the 2005 ASEE North Central Section Conference, Ada, OH, United States, 2005. 115

Pinnell, M. et al. "DETC2008–49885 Innovation, Entrepreneurship and International Experience." Paper presented at the 2008 ASME Design Engineering Technical Conference, New York City, NY, United States, 2008. 115

Purdue University. *EPICS - Engineering Projects in Community Service* http://epics.ecn. purdue.edu. Accessed January 12, 2007. 119

Riley, D. and A.H. Bloomgarden. "Learning and Service in Engineering and Global Development." International Journal for Service-Learning in Engineering, 2, no. 1 (2006): 48–59. 115

Sarin, S. "A Plan for Addressing ABET Criteria 2000 Requirements." Paper presented at the 1998 ASEE Annual Conference and Exposition, Seattle, WA, United States, 1998.

Scheibler, S. "Creating a 'Global Algorithm' for Engineering Education." Paper presented at the 2006 ASEE Annual Conference and Exposition, Chicago, IL, United States, 2006.

Stromquist, N. "Internationalization as a Response to Globalization: Radical Shifts in University Environments." Higher Education, 53 (2007): 81–105. DOI: 10.1007/s10734-005-1975-5 118, 123

Tsang, E. "Service-Learning as a Pedagogy for Engineering: Concerns and Challenges." In Projects That Matter: Concepts and Models for Service Learning in Engineering, edited by E. Tsang, 27–30. Washington, DC: AAHE, 2000. 115

Ward, K. "Addressing Academic Culture: Service Learning, Organizations and Faculty Work." New Directions for Teaching and Learning, 73 (1998): 73–80. DOI: 10.1002/tl.7309 116

Wei, J. "Engineering Education for a Post Industrial World." Technology in Society, 27 (2005): 123–132. DOI: 10.1016/j.techsoc.2005.01.001 118, 123

Zlotkowski, E. "A Service Learning Approach to Faculty Development." New Directions for Teaching and Learning, 73 (1998): 81–89. DOI: 10.1002/tl.7310 116

CHAPTER 6

International Engineering Education:
The Transition from Engineering Faculty Member to True Believer

D. Joseph Mook

INTRODUCTION, INCLUDING WHAT THIS IS ABOUT

This narrative attempts to document my "personal geography" and/or "personal trajectory" in international engineering education, as kindly invited by the organizers of this project. I believe that my early faculty career was quintessentially "typical" at a large, comprehensive, research-extensive doctoral university in the US – in my case, the State University of New York at Buffalo (UB). Following the traditional role of such faculty, I concentrated my early career efforts on research and scholarship within my technical specialty areas. I wrote research papers and proposals, was Principal Investigator on sponsored research projects, and supervised graduate student research. I also taught engineering courses, participated in professional societies, journals, and conferences in my field, and performed occasional "service" work on and off campus (committees, lectures, etc.). But mostly, I developed and maintained an active research program, which is essentially the only path to tenure and promotion within engineering – and without tenure and promotion, academic careers are quite short. I wanted a long career.

When it was time for consideration of my tenure and promotion cases, the outside evaluators dutifully wrote their opinions of my research and scholarship, and the internal evaluators dutifully read these letters and compared them with their own opinions. In the end, after two successful such cycles, I was a tenured full professor. All of this happened without any international education activity on my part, and no mention of this (lack of) activity was made by any of the internal or external reviewers.

I certainly had some international interactions early in my career, because in my research areas, as in essentially all research areas, many of the world's leading researchers happen to be located outside of the US. In fact, there were some non-US outside evaluators of my research and scholarship during the tenure and promotion review process. But these people were selected only because they were considered leaders in the relevant research areas, and not as the result of any international intent per se. Moreover, like the US evaluators, none of them commented at all (good, bad, or indifferent) about international education activity or lack thereof. It did not matter. And, of course, the research emphasis does not simply end with promotion and tenure – throughout academic careers, most of the biggest rewards and awards, exalted titles, salary advancements, etc., are based primarily on research quantity and quality.

This begs the question, why would I or any other "typical" engineering faculty member at a major US university ever get involved in international education? It is an excellent question, which is incredibly unfortunate, because by this point in our history, it should be obvious, and we should be long past the asking. But it isn't, and we aren't.

Herein, hopefully, I describe how and why I did get involved. In hindsight, it still seems a strange path to follow. As I became a tenured full professor, focused on research and scholarship in a technical specialty, I had no intentions, no prior background experience, and no formal train-ing, toward either educational administration or international education. Nevertheless, I accepted a leadership position in international engineering education, which became my primary academic specialty focus – by which I mean, over a period of several years, I found myself devoting less and less time to my technical research, and more and more time developing international activities, until, eventually, I was spending more time on the latter than on the former. I sincerely hope that this narrative can serve not only as an historical record of that transition, but also as both a "why" (or why not) and "how-to" (or how not-to) guide for others. Trends in engineering education strongly suggest that others may find themselves on a similar trajectory, sooner or later. But my story is just my story, nothing more.[1]

THE FIRE GETS LIT

I often quote the statistic that some 80% of the American population does not hold a passport. They not only have not been abroad, they apparently are not even contemplating going (at least, not anytime soon). This statistic amazes and amuses many of our international partners, and frustrates many of us trying to promote US participation. We marvel at how inwardly-focused the American general population is. Well, here is a confession: I obtained my first-ever passport for a post-college trip to Europe, and that passport expired ten years later without being used again. I never participated in study abroad as a student, and I do not even recall being aware of it.[2]

In the summer of 1992, I was invited to spend a month at the Technische Hochschule Darmstadt (THD), now called Technische Universitaet Darmstadt (TUD), to do joint research in nonlinear dynamics. This marked only the second international travel of my life, and the first in a professional capacity. By then, I was already a thirty-five year old tenured Associate Professor.

UB and TUD had a long-standing formal exchange agreement, and although I knew little and cared less about it per se, I was inadvertently participating in and benefiting from it by virtue of the outstanding TUD students who appeared on our campus and joined my research program, earning MS degrees under my supervision. I now realize that this fortunate circumstance was the direct result of a common problem in US-European academic exchanges, namely, that the interest

---

[1] In my usage herein, the term "international education" (IE) refers to educational activities undertaken for the primary purpose of simply being international. Of course, many faculty members actively participate in research activities (conferences, journals, etc.) that include non-US participants. However, the international component of those activities is essentially coincidental – the focus is the technical content. IE herein refers to research and educational activities whose primary purpose is to create international activity per se.

[2] I certainly could have done much better. My paternal grandfather was an American-born and -raised engineer who spent much of his career working abroad, in South America; my paternal grandmother was born and raised in the Alsace region of France, along the German border; she spoke the Alsatian dialect, French, German, English, and eventually Spanish. My father spent a significant fraction of his boyhood in South America, and a few summers working on his grandfather's farm in France. As a result of all this, my boyhood home had numerous artifacts and souvenirs from many faraway places. Moreover, since my father was a professor, many of our family friends were academic immigrants, including numerous parents of my best childhood friends, who often traveled back to countries of origin. Yet although I was surrounded by so many close, positive and, obvious international influences, somehow they did not lead to firsthand personal travel experiences for me until much later in my life.

from European students far exceeds that of their US counterparts. With few outgoing UB students, and an exchange agreement calling for equal numbers of students from each side, there were not nearly enough actual exchange places available for TUD students interested in coming to UB. So my department, having been very impressed with the quality of the TUD students we received, had evolved a system under which most of our incoming TUD students were actually admitted as regular graduate students with departmental financial support. Such support required that these students do a full thesis in the process of earning an MS degree, so although both UB and TUD liked to refer to the relationship as an "exchange," and a few students were actually "exchanged," most of the TUD students in my department were matriculated into degree programs and financially supported like any regular student. But I was ignorant of these details at the time, as were most of my colleagues; we were just happy to have well-qualified students as candidates as research advisees, and more importantly, we wished there were a few more of them.

After many years of this imbalance, UB and TUD administrators apparently got the idea that perhaps if some UB faculty spent time at TUD and found the experience positive, they might then be useful in recruiting more US student participants. TUD offered to financially support this effort as partial payback for the large number of TUD students that we were supporting. As the UB faculty member with the largest number of TUD student advisees, I was offered a chance to spend a month at TUD in summer 1992, and I took it.

The summer in Darmstadt by itself may have changed my focus, but what really sealed my fate was a purely coincidental encounter that occurred en route. My flight from Washington to Frankfurt was delayed, and I wandered off in Dulles airport to get something to eat. I found myself standing next to a senior colleague from UB's Department of Civil Engineering, traveling with his wife, and also heading to Germany. He was spending a few months as a visiting professor in Hanover, and he invited me to visit there from Darmstadt to give a research seminar. A couple of weeks later, I did just that. While in Hanover, I found common research interests with his faculty host, and we decided to try to find ways to collaborate in the future. I subsequently applied for and received an Alexander von Humboldt Foundation research award to support a sabbatical year in Hanover, which I took in 1994. How could I have guessed that a flight delay in 1992 would leave me in Germany for much of 1994?!?

My experience in Hanover was one of the most profound, and profoundly beneficial, of my life. Which is not to say that it was just an unbroken good time – there were plenty of challenges, down times, and bumps in the road. I went abroad in 1994 with little prior international experience, and no foreign language skills. I was separated from friends, family, students, and colleagues, living alone in a new environment a very long way from home. Yet I not only survived, but thoroughly enjoyed myself. I studied and practiced German intensively while in Germany, and I became moderately capable in a fairly short time. I traveled extensively and quickly learned how to be efficient and comfortable while doing so. I made new friends and acquaintances, Germans and non-Germans. I ate new foods, heard new sounds, saw new sights, discussed new politics and culture, and ultimately perceived the world from a completely new perspective. My broadened perspective, and awakened understanding and appreciation of another culture, permanently changed the way that I have processed world events and interpersonal relationships ever since.

I am not focusing on the purely professional aspects of my time in Hanover because I think they should be self-explanatory. After all, who does not yet "get" that the world is a huge and highly-competitive environment in which the US plays a less- and less-dominant role as time goes on? At the cutting edge of science and engineering, who is not already aware that the crucial advances are occurring around the world and that we must be engaged in order just to stay relevant, let alone important? My faculty host in Hanover was a leading figure in the field, and I recognized the importance of our collaboration in my own research, but, in fact, I could have read his papers and sent him mine from the comfort of my living room in Buffalo. The importance of being in Hanover, for me, was much less about math than it was about myth – or more precisely, about shattering many of the myths I thought I knew about myself and the world I lived in. In shattering those myths, I discovered that the truth was infinitely more interesting.

In the years since then, I have found that most students who try an international experience come home with essentially the same reaction. It is this real, life-changing, inner transformation, more than any other factor, which has motivated my efforts to contribute to international education in engineering. I want to enable and encourage as many other people as possible to have the opportunity to experience this, and given my profession as engineering educator, my efforts to do so have naturally been directed at engineering students. I believe that international experience significantly improves lives on both a professional and a personal level. That is my "why" for getting involved in international education; and I do not have a "why not." As for the "how-to", or "how not-to", well, the story is quite a bit longer and open-ended (i.e., it is still evolving and probably always will be). But details to date follow.

APPOINTMENT AS SEAS ASSISTANT DEAN FOR INTERNATIONAL EDUCATION

After returning to UB from Hanover, I became somewhat, but really only marginally, more active in helping with arrangements for incoming students from TUD, doing things such as helping with their initial arrival and finding an apartment, explaining day-to-day American life nuances in terms familiar to Germans, hosting a few parties, etc. This was entirely informal and voluntary, but it was essentially the beginning of my international education career.

Then, in early 1997, Mark Karwan, then Dean of the School of Engineering and Applied Sciences (SEAS) at UB, decided to create an official IE position and office within SEAS. Typical for him, Mark was ahead of his time; such positions were not nearly as common then as they are now. Mark offered me the position, with the title, "Assistant Dean for International Education." But with no experience and no predecessor, I had almost no idea what I might do, and to be honest, I do not think Mark did either. Nevertheless, he was serious about promoting IE within SEAS, whatever exactly that meant. I accepted the position, equally vague about it.

As a practical matter, I was given a one-course release of teaching load, representing a nominal commitment of 25% of my effort, which remained my official level of effort throughout my entire career in the position. I was provided with one month of summer salary, and I was given control of an existing Teaching Assistantship to help maintain my research program by supporting a graduate student of my choosing. I was also promised some travel and expense support, on an ad hoc request basis; the actual amount I used varied widely from year to year, but I am reasonably sure that it never exceeded about $10,000. No dedicated staff support was assigned, but I was free to ask for occasional help from the SEAS staff pool, who would try to accommodate me (and they were terrific). In total, the SEAS support, by design, mimicked a typical sponsored research program – course release to free up time during the academic year, support for a graduate student, some summer salary, and some direct travel and expense money. These terms were completely familiar to an engineering faculty member, and seemed reasonable to me at the time, especially since I had no basis for comparison anyway.

Nevertheless, I did not immediately jump enthusiastically into the effort. In fact, although I accepted the position, I agonized for years afterwards about doing so. At UB and many other US universities, most faculty administrative appointments are subject to termination at any time on the whim of the next higher administrator, creating a disincentive to throw oneself wholeheartedly into such a role. Effort and attention spent on administrative tasks is effort and attention not spent on normal research and scholarly activities, and, as noted, in a research-extensive university, faculty careers are built almost exclusively on research. Rebuilding an active research program after an extended absence can be difficult, and unless the administrative work is completely ignored, there will necessarily be a decline in the volume of research work produced during the years of administrative service, lowering career productivity. Many senior faculty members choose not to accept offers for administrative positions for exactly these reasons, and I struggled mightily with this issue during the early period of my foray into international education, often second-guessing the wisdom of the decision.

LEARNING THE BUSINESS

Early in 1997, I officially took the lead role in IE on behalf of SEAS. The position was brand new. I had no predecessor, and no specific assignment. It was up to me to create and execute a plan of action that would somehow result in increased international "whatever" within SEAS. Of course, first, I would need to figure out and evaluate the various possibilities.

The biggest early influence, by far, came from within UB. Like most major and many smaller universities, UB has a full-time university-wide office dedicated to international education. Ours is called the Office of International Education (OIE), and it is headed by UB's very long-time Vice Provost for International Education, Stephen Dunnett, who has faculty rank of tenured professor in linguistics. Within OIE work a few dozen full-time professional staff with various specialties, but Stephen is the only person with faculty rank.

The second major influence came from the Global Engineering Education Exchange (Global E3), described in more detail later. To keep the flow of the narrative more-or-less chronological, I first describe the OIE's influence. Over time, the Global E3 replaced OIE as my primary influence, but in the beginning, this was not the case since I spent most of my time on campus.

As SEAS Assistant Dean, I reported to the SEAS Dean, who reports directly to the Provost. But the OIE is, in theory, ultimately responsible for UB's international activities. OIE is headed by a Vice Provost, who is not in the chain of command above the Dean. Thus, the ultimate hierarchy of responsibility for international activities within SEAS was somewhat ambiguous. Both the Dean and Vice Provost report to the Provost, but in my tenure as Assistant Dean, I only recall one instance when the Provost was directly involved in a decision regarding an international activity for SEAS. But this ambiguity did not cause many conflicts and the three principals – Mark, Stephen, and I – were all on friendly terms and, mostly, in agreement on major issues.

The early OIE influence on me covered an enormous range of international activities. OIE naturally assumed that the creation of a formal IE officer within SEAS meant that some or all of the IE functions OIE had been performing on behalf of SEAS would now shift over to SEAS. Collectively, the dozens of full-time OIE professional staff do a lot of different things, and this expectation was never reasonable (recall, my commitment was 25%, with no support staff). But I was naïve, enthusiastic, and more-or-less presentable, so at various times I was drawn in to most of the OIE activities at some level.

Over time, I identified the following IE activities, all of which I might undertake for SEAS. In the next sections, I discuss these in more detail; here, I simply list them for convenience:

1. Create, negotiate, and/or manage international academic partnerships for SEAS and/or UB (usually bilateral agreements with a single international partner). I include herein visits to, and hosting visits from, international partners and potential partners

2. Strategic planning / advising of administration (Chairs, Deans, Provost, President)

3. Plan, launch, and manage international branch campuses and off-campus programs (Here, I refer not to programs established as study abroad for existing students from the home campus, but to those established in order to enroll additional international students who may never set foot on the home campus.)

4. Recruit and/or support incoming international full-time, degree-seeking students

5. Seek outside funding to support international activities

6. Participate in IE professional organizations, conferences, journals, etc.

7. Participate in US academic consortia engaged in group IE activities; may include representing them in relationships with non-US consortia, government, etc.

8. Speak, research, publish, advocate, publicize, and proselytize on behalf of IE in a wide variety of forums, on and off campus

9. Create meaningful, successful study abroad programs for students

10. Recruit, enable, and encourage student/faculty participation in international activities

11. Support incoming students, faculty, and scholars from exchange partners

12. Encourage and educate parents, faculty, and others to support outgoing students

I want to strongly reiterate that this list of potential activities is the result of a long period of learning and of observing others. I eventually did all of these to some extent, but I could not have even written this list in the beginning. All of these activities are important, and many could, by themselves, easily and productively utilize a multi-person full-time staff. But I was a "staff" of 25% of one person! My challenge was to figure out what I should do and how I should do it within that severe constraint.

Note that there are essentially three main groups of activities in the list. To make an analogy with traditional engineering faculty roles, items 9-12 are IE versions of mainstream faculty activities such as teaching, research, and other direct engagement with students and colleagues; items 5-8 represent IE scholarly activities like presenting and/or publishing results and engaging in professional interactions with others in the specialty field. Thus, from the perspective of a faculty member focused on an IE specialty, items 5-12 are the top priority. Items 1-4 are more appropriate for IE staff or administrators. I was both "IE" faculty and administrator.

The influence from Global E3 (and several other sources as time went on) pushed me mostly in the direction of items 5-12. In stark contrast, the priority within UB's OIE, and, I imagine, the equivalent IE offices at many other US universities, is essentially on items 1-4.

WHAT I ULTIMATELY DID, AND WHY

With the benefit of hindsight, I attempt to describe in more detail what I did (or did not do) with respect to each of the items listed above.

1. *Create, Negotiate, and/or Manage International Academic Partnerships for SEAS and/or UB (Usually Bilateral Agreements with a Single International Partner):* At UB, as at most universities, these agreements are normally enacted via a Memorandum of Understanding (MOU), a formal document which must be drafted, edited/approved, and finally signed by appropriate officials from both sides. Existing MOU's are also subject to review, possible modification, and then renewal (or cancellation), typically every five years. The actual writing, editing, and approval process, whether new or renewal, can be tedious and time-consuming, but it is critically important to get it right.

   Partnerships, and potential partnerships, also normally include official visits in both directions, typically consisting of an intense schedule of meetings, meals, sightseeing and/or entertainment that can last several days, from breakfast until after dinner, and that does not include travel time and/or planning and preparation time.

When I got involved in IE, I was more or less completely oblivious to just how big, and important, this activity is, and frankly, I think this is vastly underestimated by most of those who have not actually had this responsibility themselves. There were probably years in which my hosting responsibilities alone, if all hours were counted, exceeded my entire 25% IE commitment. A single visit, including the planning and communication ahead of time, can consume most or all of an entire work week, and there were years when we hosted a lot of these visits! The situation is different for an individual faculty member conducting only his/her own program with one visit per year, but I was called upon to represent SEAS and/or UB whenever any partnership might include any aspect of engineering, and even quite a few which did not.

I thoroughly enjoy the personal and professional interactions with our international partners. It is wonderful to meet like-minded people and ultimately craft a successful partnership that will provide life-changing educational opportunities to students and colleagues. The enormous time and effort spent is unavoidable, so it is good that it was actually a highlight of my experience. Some of these partners have become close personal friends outside of our professional obligations.

Many academic institutions boast about the number of formal agreements they have in place, in much the same way that faculty members quote numbers of publications – in both cases, it is assumed that more is better. In reality, many of these formal partnerships essentially exist only on paper. We try to limit formal partnerships to those in which some real activity exists, or can be reasonably expected, and furthermore, to those in which a formal agreement would provide some actual tangible benefit. Plenty of international scholarly collaboration can and does occur without the existence of a formal agreement between institutions.

The most common practical benefit of formal agreements is the establishment of a student exchange mechanism. Normally, exchange students pay tuition to their home institutions, but they attend the partner institution, which does not charge them additional tuition. At UB, we must have a formal agreement in place, after which the accounting is very simple – one incoming full-time semester student for each outgoing full-time semester student, regardless of major or year of study. This system existed for a long time before my appointment, but at some point, I realized that I could use it to generate new graduate student tuition waivers as a financial benefit of our study abroad programs. I did this by creating undergraduate study abroad opportunities for UB students with international partners from whom we routinely recruited and supported incoming graduate students. Now, UB students who participate in those programs effectively produce tuition waivers for additional incoming graduate students from those institutions. I used this model in crafting several programs, resulting in additional graduate enrollment at UB worth hundreds of thousands of dollars of tuition. The economic side-benefit was very important, but in creating study abroad, my primary motivation was always to create a high-quality academic experience for UB engineering students, and I describe some of those programs below.

Over the years, I have traveled often on behalf of SEAS or UB with respect to formal bilateral partnerships. In the beginning, I usually visited existing partners. Since so much rests on the personal relationship between officials in international relations, it was important for me to become acquainted with my counterparts. In addition, my visit was sometimes part of the process of renewing an existing partnership. But in time, the most common reason was to find ways to encourage UB student participation, since, like many US schools, recruitment of outgoing students is the bottleneck in the exchange. These discussions could be delicate since it put me in the position of evaluating the partner and suggesting changes or additions solely for the benefit of our students. However, I was always careful to make it clear that my motivation was to increase student participation in both directions, which necessarily required that we get more UB students involved, and I do not recall any instance where this good intent was misinterpreted.

On several occasions, under two administrations, I was asked to join the UB President as part of his delegation during official overseas visits, which were always organized and escorted by the Vice Provost. These types of trips normally included a lot of high-level ceremony, often with very elaborate and generous hospitality, and it was both interesting and important to participate. On a personal level, both presidents (and their wives) were delightful traveling companions. The presidents themselves would not normally engage in most highly detailed discussions regarding the particulars of the partnership, but behind the scenes, I often worked closely with the Vice Provost on such details.

2. *Strategic Planning / Advising of University Leaders (Chairs, Deans, Provost, President):* Good academic leaders will solicit and carefully consider advice from experienced, well-intentioned sources around campus, but not all academic leaders are secure or savvy enough to do so. Mark Karwan established the Assistant Dean position for a purpose, and he expected me to become the expert within SEAS. He routinely sought my advice and was always willing to carefully consider my strategic proposals for new initiatives.

I had the opportunity to participate in several formal campus-wide planning exercises, the largest one of which occurred during Spring 2007 when I was a member of a Provost's Task Group charged with writing a strategic plan for internationalizing the campus. We produced, in my opinion, a very good and detailed plan for UB that was warmly received by the provost and president. It remains to be seen how much of it can be implemented. One of our key recommendations was that every decanal unit at UB should have its own equivalent to an Assistant Dean for International Education. This turned out to be a bit ironic given what would happen to that position, my position, within SEAS at about the same time, which I discuss in the final section.

3. *Plan, Launch, and Manage International Branch Campuses and Off-Campus Programs:* At the time that I became Assistant Dean in SEAS, there was only one other School at UB—the School of Management (SOM)—that maintained a separate office for international education

outside of the University-wide OIE. The SOM's primary international education activity was to offer MBA or executive MBA (EMBA) programs at various locations around the world. In fact, this activity is common to a lot of American business schools, and it has been lucrative for some of them, including ours. In a typical scenario, a sequence of UB professors provides short, intensive versions of MBA courses, commonly two or three weeks in length. These courses are offered sequentially, over a period of perhaps a year. The students, many of whom are professionally employed, take one course at a time; faculty absence from the main campus in Buffalo is manageable for the short duration.[3] The program is sold as a full-degree package with a price that enables the individual faculty members to be well compensated, and at the same time, it generates additional resources for the school. The SOM Dean and the Provost both get a big cut.

UB's SEAS operated a branch campus in Kuala Lumpur for several years, but that activity ended before my appointment. This campus provided the nominal equivalent of the first two years of UB engineering programs, at a fraction of the cost, using locally-hired instructors rather than UB faculty, and students were then encouraged to transfer to UB for the final two years to earn a degree. The operation was apparently lucrative for some time, aided by Malaysian government support, but it was eliminated before I had any involvement with it. Needless to say, the SEAS Dean and the Provost shared in the profits, and the Vice Provost managed the operation.

Against this background, over the years, usually with the strong urging and support of the OIE, I investigated various potential versions of similar off-campus programs in engineering. The motivation was invariably to generate enough revenue to pay expenses and then create sufficient resources back home to justify the effort. I designed and eventually proposed several versions at a number of different levels – two-year starter programs, which would feed transfer students to UB; full four-year BS degree programs; and SOM-style MS or MEng programs. However, in the end, none of these programs were offered. Although the reasons varied by location and program, most programs could not be made cost-effective enough to attract the support of the SEAS Dean and Provost. I never proposed a program that was not self-supporting and academically rigorous, but these were not good enough reasons to move ahead unless the profit margins were higher.

4. *Recruit and/or Support Incoming International Full-Time, Degree-Seeking Students:* Many academics in international education might find it odd that I include this in my discussion. But this is by far the biggest activity within UB's OIE, and OIE was my early influence. Soon after my appointment as Assistant Dean in SEAS, OIE arranged for me to join a month-long recruiting trip through eight Eastern Asian nations, attending dozens of educational fairs accompanying our head international recruiter, Joseph Hindrawan. We were recruiting for the

---

[3] As an aside, to the best of my knowledge, the SOM did not, and still does not, offer any study abroad opportunities for home-campus UB students.

incoming Fall 1998 class, and as I understand it, numbers of international applications in engineering were actually quite a bit higher that year.

This was my only such trip, however. It is not that I do not think it is important for UB. We benefit tremendously financially, as well as academically, from international students. In engineering and sciences, international students comprise more than 50% of our graduate enrollments, in some departments, more than 80%. The OIE dedicates more time and effort to this function than any other international activity, by far. And SEAS has its own enroll-ment management effort, and moreover, each Department has a separate graduate admissions activity. I decided that my part-time effort should be spent on other activities, specifically, those that included the recruiting and sending of UB engineering students into international programs.

5. *Seek Outside Funding to Support International Activities:* For many faculty from engineering schools, external financial support is inextricably linked to most professional activities. It is in our professional culture that we assume that most things worth doing are, by definition, worthy of support from some external sponsor. Although I was reasonably well-funded in my research activities prior to taking on the title of Assistant Dean, I have spent very little time and energy seeking external funding for international activities. There is no particular reason for this, except that I have always looked at sponsored funding as a tool, not an end in itself. External funding is normally supposed to be seed money only, and most of the programs that I established were designed to be self-sufficient from the start. This may sound blasphemous at a major research university, but the truth is, I did not really need external money, so I did not work very hard to find it.

6. *Participate in IE Professional Organizations, Conferences, Journals, etc.:* I was a very active par-ticipant in such activities in my technical research, but it was still a pleasant and productive surprise to join and participate in appropriate professional international education organi-zations, conferences, consortia, etc. For everyone, but especially for novices, these activities provide access to a wealth of experience and ideas. International education can mean so many different things at so many different institutions, and it is a constantly evolving field, so it is both important and incredibly interesting to maintain close contacts with others who are active. Last but not least, the chance to spend a few days surrounded by people who share our passion is terrifically motivational.

In my own case, by far the best overall experience has come through the Global Engineering Education Exchange (Global E3), which is described in the next section. I also benefited tremendously from attending the Colloquia organized by John Grandin at the University of Rhode Island (more recently held in other locations as well). Larger venues, including NAFSA, the ASEE Annual Meeting, and Frontiers in Education, among others, all have their good points, but I still prefer the smaller conferences and symposia that are completely dedicated to engineering international education.

7. *Participate in US Academic Consortia Engaged in Group IE Activities:* UB was already a member of the then newly-formed Global Engineering Education Exchange in 1997. The Global E3 is an organization of US universities facilitating engineering student exchanges among about 100 universities in about twenty nations. European partners are grouped into their own formal organization, based in Paris and now called GE4. In the beginning, the focus was on American-European exchanges, but over the years, additional universities, primarily in Asia, have been added. The administrative structure has also changed somewhat, but the focus remains on true academic exchanges of engineering students, nothing more and nothing less. Global E3 and GE4 – the similar names are pure coincidence - take turns hosting an annual meeting. The organization and annual meeting are entirely dedicated to engineering exchanges, and thus it is an incredibly fertile ground for exploring ideas, learning from others, establishing important contacts, and launching new program initiatives. Moreover, the program always includes cultural, social, and professional outings that enable participants to really get to know one another – a critical element in forming good international partnerships.

   The tangible benefits of membership in this consortium are many. With so many partner institutions, an engineering student in just about any major and any year of study can find a suitable placement; more likely, there will be several to choose from. The US students at any particular overseas location often come from a mix of US member schools, widely varied in terms of location, size, and focus, enabling US students to mingle in ways that never occur back home. In many cases, the host institution can help to arrange an internship to follow the academic semester or two, so the entire experience can last a year and include both academic and work experience. Institutional membership in Global E3 is based on a single MOU with the organization, and a member's student exchange balance need only be maintained in total across the organization, but not necessarily with any of the individual member institutions. This is a truly incredible benefit, eliminating an enormous administrative overhead.

   It was my honor to be elected to the Global E3 Executive Committee in 2001, elected its Chair in 2003, and re-elected as Chair in 2006, a position that I still hold. However, Global E3 is certainly not the only consortium; in fact, I have joined, and even organized, a number of smaller, ad hoc consortia for targeted purposes. In every case, the primary benefit is that the additional US students help to ensure sufficient participation to justify the program and spread the real risk of low US student participation across the partners.

8. *Speak, Research, Publish, Advocate, Publicize, and Proselytize on Behalf of IE in a Wide Variety of Forums, on and Off Campus:* As the SEAS official in charge of our international programs, I spoke about IE frequently on and off campus. The format, focus, and audience varied widely. At one extreme, I made many presentations to high schools and to potential UB students and their parents during UB Open Houses (attended by high school seniors undecided about which college to attend) and Preview Days (essentially an orientation day for high school seniors and transfer students who had made the decision to attend UB). I also normally made a presentation to new engineering freshmen on Opening Day (which is actually the day before

the first day of class in the fall semester). Later in the fall semester, I gave each of the several freshman Introduction to Engineering course sections an entire lecture class (fifty minutes). Throughout the year, I would give a number of smaller, targeted information sessions for our various engineering study abroad programs.

At the other extreme, I would summarize our activities for outside groups such as the engineering Dean's Council (an advisory board), local citizen's and business groups, and government officials. I also made occasional presentations to various groups of faculty and/or administrators on campus. In addition to these presentations, I frequently provided quotes, written information, and even draft articles for various on-campus publications.

9. *Create Meaningful, Successful Study Abroad Programs for Students:* In the beginning of my international education career, I do not remember giving this activity any thought at all, and yet today this drives most of what I do in IE. Although UB was already a member of Global E3, this is not the same as a study abroad program set up specifically for US students in some international location, which may or may not have anything to do with a partner institution. Even when it does, a special study abroad program is not as integrative as a pure exchange.

Study abroad programs at UB – outside of engineering – are mostly the result of a specific personal interest on the part of the faculty member. For instance, a professor of European art history might establish a study abroad opportunity for students to study art in Europe. This motivation remains largely absent in most engineering programs, for the simple reason that the course content in most engineering programs is simply not location specific, e.g., there's no reason that one must go anywhere in particular to study calculus or thermodynamics.

It is my impression, perhaps incorrect, that the most common format of engineering study abroad at most US schools is to send a group of students, accompanied by a professor, to live and study as a group at some international location. The duration of many of these programs is much less than a regular semester.[4] By nearly every measure, such programs are a significantly less intense experience for the student than a pure exchange that immerses him/her completely in the regular life of the host students for one or two full semesters, taking their courses, speaking their language, living in their dorms or apartments.

Thus, I consider pure exchange to be the ideal – an ideal that is strongly emphasized in Global E3. Students just trade places and take existing courses, eliminating much of the special effort required by faculty and administrators. But the intensity of this experience, not the least of which is the potential language requirement, is an impediment to large numbers of US students, and participation remains frustratingly low.

In an effort to address the problem of low participation, I created a few engineering study abroad programs that fall somewhere between the cocoon and the ideal. These programs have

---

[4]I know of some as short as two weeks, which suggests the following practical questions: Is there a duration so short that it is essentially academically meaningless for the student? What is the minimum program that is worth the time and effort of the faculty and/or staff who must organize it?

had the intended effect of attracting far more students than were previously participating in full exchanges, and some students on these lesser programs go on to a full exchange experience later.

When designing a study abroad program, I believe that the academic program should provide full, usable credit towards a degree, and if possible, should include academic portions such as history and language, that are directly related to the location of the program. This latter part is often the biggest challenge at UB because within our engineering degree programs, the number of credit hours allotted to these general education courses is quite limited.

I first established a five-week summer program in France, hosted by the University of Technology in Troyes (UTT). The academic portion fits students who have just completed the freshman year, and the intent is not only to give them this international experience, but also to motivate them for another one before graduation. UB students earn six credits, all usable toward graduation – three for statics, and three for a course called World Civilization II. This latter course is required of all SUNY students, but the exact content can vary from section to section within guidelines. The students live and study as a group, and a UB professor teaches the statics course, but other features are designed to maximize integration within the constraints of a summer program. The program takes place before the end of the semester in Troyes, so the UTT students are still in full residence. There are a number of planned activities that bring UTT and UB students together outside of the classroom, including a three-day "Adventure Weekend" in the south of France that is a huge hit. The World Civilization course is taught by UTT faculty, focusing on EU and French issues, bringing in a local flavor. Students with some French language background can also take intensive French courses. The city of Troyes is manageable but interesting (140,000 residents and a long history), challenging and enabling UB students to be adventuresome around town. There is a weekend in Paris and a free weekend during which most UB students scatter off to other European cities to sightsee. The program is far from the ideal exchange experience, but for five weeks, the students get a pretty good taste of France. It is not exactly cheap, but the cost reflects tuition and reasonable expenses for the time in France.

For every three UB students who participate, a full-semester tuition waiver is generated at UB. There is strong interest among UTT students for pursuing MS degrees at UB, and UB has strong interest in receiving them. The waivers from the program support such students, now numbering in the dozens over the life of the relationship. Some of the initial MS students so impressed their UB faculty advisors that they have since entered UB Ph.D. programs. Thus, looking at the big picture: Young (freshman) UB engineering students get an international experience, earn suitable credits toward graduation, and have plenty of time to follow up with another study abroad later in their programs (many do this); UTT students get tuition waivers and an opportunity to earn an MS degree at UB; UB Departments get additional graduate students supported from outside the Department. Everyone benefits, and the program is self-supporting and sustainable.

I have since followed up the program in Troyes with two junior-level "targeted semester" programs. "Targeted" means that they are specific to certain degree programs. In Toulouse, home of Airbus and several long-standing UB partner schools, the program is targeted to Aerospace Engineering but is also suitable for Mechanical Engineering. At ENSEA, outside Paris, the program is targeted to Electrical Engineering. In both places, courses are taught by French faculty in English. At ENSEA all courses are set up specifically for the US students, but in Toulouse, about half of the courseload is identical to locally-required courses (except that it is taught in English), and those courses are mixed with local students.

To justify the effort and expense of establishing the English-language courses, enrollment targets for US students were set in each location. In anticipation that UB alone would not meet this target, a select group of US partners also sends students into these program – mini-consortia of sorts. This required additional background work to establish suitable course content to satisfy the requirements of each US school, a challenging but not insurmountable task. The US students in the program thus mingle with each other as well as with their French counterparts, providing an additional cultural bonus.

Students spend a full semester, live in regular dorms randomly interspersed with French students (dorm rooms are singles for all students), and follow the normal semester schedules of the local students, maximizing integration to the extent possible without requiring fluency in French. Most of them see other US students in their classes but tend to socialize with their dorm friends and neighbors. During the semester breaks, they scatter around Europe and North Africa as tourists. Whether they like it or not, they typically learn some French if they do not know some prior to joining. Several students in Toulouse have stayed on for a full year, doing a local internship at the end of the academic semester. One of these worked on the A380 team through the period when critical flaws in the wiring were detected, leading to the resignation of top corporate officers – a major international story, and certainly one of the leading aerospace engineering experiences possible in the world at the time.

My creation of all of these programs helped me get over some of my initial reluctance to throw myself into international education because they have many very desirable features for our research-dominated academic culture. They produce substantial support (the tuition waivers) for additional graduate students, mimicking one of the key benefits of sponsored research. The incoming students have proven to be of very high quality and are in high demand among UB faculty, another research-like benefit as well as a key academic benefit within our graduate program. And, as noted, a number of them have subsequently entered our Ph.D. program, the "ultimate" prize. Likewise, the UB students who participate are about five times as likely to be Honors students at UB, and twice as likely to be female, so the programs appeal to domestic student demographics of high interest to us.

All three of these programs are self-supporting and sustainable, but due to the UB Department's financial value of the waivers for incoming students, several are now offering small travel scholarships to UB students to encourage additional participation. In dollar terms, the payback is about 10 to 1, based on out-of-state graduate tuition that would otherwise be paid to the incoming students. So there is yet another benefit for all concerned! We have several other programs, but I've chosen to cut myself off here.

One final note: The process of trying to include locally-relevant courses such as history, language, politics, etc., is a terrific vehicle for expanding your own understanding of the local culture.

10. *Recruit, Enable, and Encourage Students to Participate in International Activities:* Statistics indicate that even today, only approximately 2% of US engineering students participate in international experiences at any time during their entire degree program. As with so many other aspects of IE, I was essentially oblivious to this fact when I started – I thought most students would be thrilled with this possibility. In fact, it is not clear that most of them ever seriously consider it.

In truth, I still do not really understand why this is, but I did manage to raise UB's participation rate to between 10-15%. I emphasize that cost is probably not an issue; academic credit can almost certainly be arranged; many locations teach in English, if that is an issue (it usually is); most importantly, 96% of the world's population lives outside of the US, it is a global economy, and the future lies in preparing for it.[5]

My experiences with recruiting are mostly based on UB students, and, of course, the demographics at every university are somewhat different. I find that successfully recruiting students requires two complementary components. One component should emphasize the positive aspects of the international experience: previous student participants, with their pictures, can be excellent in this regard. I also think it is important that a faculty member emphasizes the positive educational and professional career aspects of the experience. The other component is to preemptively guess and address any concerns that they have, i.e., negate the negative. Waiting for questions is too risky, so I bring these up myself, then address them: cost, credit, language, housing, homesickness, and potential animosity towards Americans.

11. *Support Incoming Students, Faculty, and Scholars from Exchange Partners:* The bulk of this paperwork was usually done by staff in OIE, but I did successfully change the procedure used for incoming exchange students to ease their application process. In the beginning of my term, exchange students were processed into UB using mostly identical criteria and procedures as degree-seeking international students. This meant that they had to provide original transcripts, proof of English skill via a minimum TOEFL score, etc. However, these requirements seemed unnecessary to me because exchange students are non-matriculated, and UB is not risking its reputation by awarding any degrees. Moreover, most of our partners readily accepted any student that we sent, without question. In effect, exchange students are guests, and I felt that our partners were treating our students much better than we were treating theirs. I began to

---

[5]To make this point, I used to say, "You can't win Wimbledon if you only practice tennis with your kid sister in the backyard." One day a student responded, "You can if you're Venus Williams." I don't use that analogy any more.

push for relaxing the process and removing some of the barriers. This met with stiff opposition within OIE, for reasons that never made sense to me. My perspective was eventually supported by the Director of Study Abroad at UB, and together we convinced the Vice Provost to give her the responsibility for incoming exchange students since she was already responsible for outgoing exchange students and thus in the forefront.

Over the years, I also found myself spending a fair amount of time and effort helping with all manner of arrangements associated with visiting scholars and faculty in SEAS. Many of these details are pretty mundane although important for the visitor, and staff support, when I could get it, was greatly appreciated. An arriving faculty member, perhaps accustomed to a certain amount of deferential treatment, and perhaps traveling abroad for the first time, can be pretty demanding.

12. *Encourage and Educate Parents, Faculty, Others to Support Outgoing Students:* This was mostly fun, because I believe so strongly in the cause, and because most of the people I dealt with were already enthusiastic about helping. It can get time-consuming, however, especially with those parents who want personal answers to all their questions. I will simply share one funny anecdote about a mother who called me several times prior to her son's participation in one of our study abroad programs. I established the program as an easy first experience abroad, relatively short-term, in a group of UB students, and escorted by a UB faculty member (myself, in this instance). Nevertheless, she was clearly stressed about the concept of her son going abroad, calling me repeatedly with new questions, so I finally tried to assure her by emphasizing that the trip was designed to accommodate less-experienced travelers. At this, she indignantly pronounced that he was an experienced traveler – the family had once driven to Dayton. I apologized for my mistake; the son joined the trip, looked a little green at the airport, but ultimately had a great time.

SUMMARY AND CONCLUSIONS

International education in engineering, though growing rapidly in importance, has still not yet reached a point where new engineering faculty members can launch and sustain a successful career with it as their primary specialization. If an engineering faculty member is to become heavily involved in international education activities, it will almost certainly be the result of some mid-career transformation of focus away from traditional research activities and into less well-traveled and understood educational activities. This will cause discomfort to most faculty members, concerned about the long-term career implications. I have tried to capture the essence of my transition from research-driven, IE-oblivious faculty member, to a leadership position in IE. I hope that my descriptions along the way are sufficient to understand why and how I did what I did.

In the end, I focused on establishing study abroad programs, within exchanges, as the number one priority, but this necessarily included a number of the other activities. I think it would be quite possible for an engineering faculty member working in a university with good staff support, especially good support in the university international office but perhaps even only with standard engineering staff support, to pursue these kinds of programs without the need to do so much of the administrative work him- or herself.

In doing this work, I was able to increase UB engineering student participation by between five and ten times the previous rate and provide these students with academic and personal experiences that transformed their lives in positive ways. Moreover, through the exchange mechanisms of the programs, I simultaneously increased the number of high-quality incoming graduate students in engineering at UB, and I also found a way to make the financial underpinnings of these programs substantially valuable to UB while simultaneously being self-supporting. In short, in about the same time that it took me to establish the successful research programs that carried me through tenure and promotion to full professor, I managed to create unique international programs that provided truly outstanding educational opportunities, increased graduate enrollments, provided substantial and important financial benefits to the university, and increased our visibility and reputation among peer institutions around the world. By most reasonable measures, my international programs produced tangible academic results that met or exceeded those of many research-active senior faculty in engineering. So although I started down the path from research-active faculty member to international education with many doubts about the long-term wisdom of the decision, in the end, I look back with pride and no regrets. And I have the satisfaction of knowing that the impact of these programs on students far exceeds the impacts of most of their other experiences at university, and this statement is proven repeatedly by exit interviews with graduating seniors. Those who participate in study abroad overwhelmingly identify the experience as the greatest of their undergraduate careers.

Nevertheless, without so much as a conversation, let alone a formal review, the position of Assistant Dean was eliminated on the whim of a new Dean. I do not think this outcome is likely to be repeated at most other universities, even those under mediocre or worse leadership. UB's inexplicable leap backwards notwithstanding, across the US, we are rapidly reaching the point where not having an international program at a major engineering university might be compared with not having a library. Accordingly, I am confident that a growing number of engineering faculty at major research universities will find themselves becoming involved in international education in the future, and so I end with a short list of advice:

- Ask for a detailed description of what is expected of you, along with a detailed description of how your performance will be evaluated. Do not rely on the good will of the person who appoints you, as that person may be replaced by someone with a different viewpoint.

- Using the range of international activities possible, as described elsewhere in this article and/or in other articles, ask for a philosophical and/or actual ranking by importance to the institution. If you disagree with this ranking, do not take the position.

- If other official positions with international responsibilities exist on your campus, ask for a detailed description of the hierarchy and working relationships between these offices and yours.

- Insist on a formal, regular, and objective evaluation procedure, conducted with input from outsiders who can be counted upon to act in good faith. Ask for a commitment describing how the results of this evaluation will be used, if at all, especially as it relates to discretionary salary increases, promotions, etc. You may be the only person in engineering doing this work, and the discretionary system may not know how to handle that.

- Insist upon some staff support, which might be as little as some hourly-wage student help. There are a lot of mundane tasks that you will quickly tire of.

- Insist upon reasonable travel support. At a minimum, this should cover attendance at two or three professional meetings per year, and in addition, one or two trips abroad to visit existing and/or potential partners. Use these trips wisely in the early stages by attending focused conferences and meetings such as the Global Engineering Education Exchange.

CHAPTER 7

Finding and Educating Self and Others Across Multiple Domains:
Crossing Cultures, Disciplines, Research Modalities, and Scales

Anu Ramaswami

LEARNING HOLISTICALLY ABOUT NATURE, PEOPLE AND TECHNOLOGY IN INDIA

I was born in New Delhi, India, where I lived for the first fifteen years of my life. At that time, New Delhi still had many parks and open spaces and its own abundant "urban wildlife."[1] I would wake to the caw-cawing of beautiful peacocks that often roamed in our backyard, and harmless lizards, all given nicknames by us kids, crawled beneath the tube-style florescent lamps on our walls. There was also the usual melee of stray dogs, cats, cows, mice, and even the occasional mongoose; fox were known to come in from the open space to rampage; scorpions, centipedes, and leeches were to be watched out for in the rainy season. As kids, we accepted and navigated around these creatures while we got to know about the local plants and trees - two guava trees in our front yard, a custard apple tree and a drumstick tree (moringa) in the back yard replete with fragrant jasmine bushes, a mulberry tree (of silkworm fame) across the street, flowering tall cottonwoods that lined the street, and various berry bushes (Jamuns) were everywhere.

My childhood was largely spent outdoors with games involving nature. While I am sure I know much less about nature compared to children who live on rural farmlands, I look back at my childhood years and recall a strong, deep, and spontaneous holistic connection with nature. It was a connection created without the formal, practiced use of nature walks, camping trips, hiking trails, and other devices that are now used to connect urban children to nature. This connection stayed with me and influenced me later in life.

All the senior women in my home (mother, grandmothers, and an array of aunts) and all the maids who worked in our home over the years (maids were common in even the most modest homes in India at the time) provided immense knowledge about living and connecting with nature in a healthy and holistic way. Their knowledge ranged from common-sense approaches to keeping food fresh and safe from ants to optimally scheduling the staggered harvesting and ripening of fruits to reduce spoilage (we got mangoes, coconuts, and custard apples from our trees) and to specific knowledge on the poisonous and medicinal properties of various plants. I grew up respecting this knowledge and wisdom, particularly because I saw it work time after time, most importantly, as a cure or a relief to various illnesses. *Neem* tree leaves in bathwater soothed the itchiness of childhood chicken pox. *Fenugreek* seeds were an effective remedy for the "stomach virus." Eating cooling vegetables and foods like raw onions prevented sun stroke, etc. The most impressive example was the use of a certain herb (*kizhanelli; Phyllantghus Amarus*[2]) to treat jaundice (Hepatitis) because we

---

[1] This vocabulary of a *separate* wildness – separating all other life from humans - is a uniquely Western concept that I have now learned after living many years in the US. In India, all life was accepted as such.

[2] Jayaram et al., "Efficacy of Phyllanthus Amarus Treatment in Acute Viral Hepatitis A, B and non A non B," 1997.

could see the yellow tint in my sister's jaundiced eyes and skin fading away as she used the juice of this medicinal plant.

The science of nature was also applied holistically by women to increase the comfort of our homes. These techniques ranged from evaporative cooling of water in *surais* to air-cooling using *Khus-Khus* coolers, fuel conservation while cooking, and a range of inter-connected strategies to keep warm in winter. What was impressive about all of their knowledge and wisdom is that the senior women in my home - my teachers - not only knew much about these nature-based techniques, they were also equally adept in handling, running, and maintaining more complex gizmos such as kerosene stoves, butane gas cylinders, pressure cookers, and sewing machines. My mother would routinely provide precise and effective instruction to my cousin brothers on how to jack-up the car to change out a flat tire!

In retrospect, I view such women's knowledge as quintessential holistic engineering: my mother, aunts, and grandmother used creativity and knowledge to solve practical problems in order to increase human well-being day after day after day. In their efforts, they considered the integrated human-natural system as whole. They practiced food science and engineering, biological engineering, and a bit of mechanical engineering and HVAC engineering every day. In their practice, while they were not technical experts, they had a strong connection with the user sphere of technology[3] that made them strong *engineering technicians*. Combine women's practical everyday use of technology with their everyday experience in supply chain management, just-in-time scheduling, and strategic labor management from routinely organizing feasts and associated religious functions attended by fifty to100 family members, and one could see that the senior women in my home had strong hands-on training to be the best of engineers and managers.

The knowledge and wisdom I learned from my father, uncles, and grandfathers, in contrast, was more in the plane of theory and strategy, representing the virtuosity values of the "expert."[4] Coming from a family of mathematicians on my father's side, I loved mathematics and was expected to excel in that subject.[5] Growing up, our group of ten-plus cousins – boys and girls - would often play card games, board games, and also team sports (badminton, tennis, and tennis-ball cricket) over summer vacation. Winning in games was a gender-neutral trait in our home. The senior men in our family participated actively in these games, which became a learning ground for strategic thinking. My father, in particular, had a penchant for critically reviewing each game to understand errors to be corrected. At the same time, he was the most positive person with a "can do" outlook toward the future, who firmly believed anyone could do anything they set their mind to – gender

---

[3] After Pacey, *The Culture of Technology*, 1983. In mapping the medical professions, Pacey maps nurses to the user sphere in medicine (nurses embody user-need values) while medical doctors are mapped to the expert sphere (doctors exemplify associated virtuosity values).

[4] Here once again, I am following the terminology of expertise or virtuosity values coined by Pacey.

[5] I noticed that in many other well-educated South Indian families, there was an equal expectation for young girls to excel in mathematics and science just as much as young boys. While young boys may have experienced greater pressure to excel in these subjects in order to become wage earners, I, as a young girl, experienced no negative commentary on the "inherent" ability of women to succeed in math of science as I have come across anecdotally in the US (e.g., Lawrence Summers, Barbie Dolls, etc.).

being irrelevant. This triad - strategic evaluation, continuous improvement, and a basic confidence in achieving a goal - are important aspects of my engineering professional life in the present.

Thus, although I did not realize it at the time, both the male and female influences during my school-going years helped shape my future thinking in sustainability engineering. Indeed, in a finite world, sustainability engineering requires connecting the "user sphere," or the user demand-side of technologies for water and energy, with the "expertise sphere," which is traditionally focused on the supply side. In the sustainability literature, these two spheres are also referred to as the soft path and the hard path.[6]

Although I went on to attend one of India's best engineering schools for an undergraduate degree in chemical engineering, my college experiences were curiously not as foundational in establishing the linkage between "user-need" and "expert-virtuosity" engineering values, nor the connection between nature and people, as were the childhood experiences described above. At the college level, I encountered a strong gender imbalance in the engineering student population. At the Indian Institute of Technology that I attended, the percentage of women among engineering undergraduates was only 1%. The curriculum was rich in theory and calculus – all of which I enjoyed and recognized as important– but it did not provide hands-on practical open-ended engineering experiences. The social and cultural aspects of engineering and connections with natural systems were not mentioned very often, and the values and assumptions underlying engineering were rarely challenged at the time (mid 1980s). The primary values and assumptions that were on my mind even then were:

- The role of engineers as agents of social change not as passive providers of technology[7].

- Related to the above, discussing the complexities at the junction of technologies (supply side) with people (user sphere).

- The ethical role of the engineer in assessing the "good" and "bad" impacts of technology, and wondering who defines these attributes.

- The now-fashionable view of engineering as being sustainable and symbiotic with the natural ecosystem, versus another view that projected engineering (often with masculine connotations[8]) as an activity that "harnesses" and "controls" nature for human "use," a view that indirectly commoditizes nature.

[6] For hard and soft paths in the energy sector, see Lovins, "Energy Strategy," 1976. For the water sector, see Gleick, "Global Freshwater Resources," 2003.

[7] Several of us students at IIT worked toward social change in our own small informal ways, in addition to having a program called National Social Service. Informally, I was most involved with the lives of the workers who worked in our dormitories and the plight of stray dogs. Not all these projects ended well. Although against rules, I fed and raised several stray dogs in our dormitory, many in very sad physical condition, one of whom had distemper and died. A more complex project involved trying to move our dormitory janitor's daughter and son (both of whom were deaf and dumb) to a vocational school for the deaf and dumb. In the end, strict cultural mores of the day prevailed and this young teenage girl was married to her uncle (as was the local custom in that community), who was also deaf and dumb, and abusive, in addition. I learned directly how difficult it is to effect social change even at the individual level, and also that cultural beliefs exist and must be challenged internally (not externally by me, the outsider).

[8] After the works of Vandana Shiva, Carolyn Merchant, and Jerry Mander.

Stimulating conversations addressing these facets occurred independently outside of class as discussions among peer groups of students as we would share our developing ideals, ethics, and aspirations. In theory and computational classes, I experienced almost no gender bias on a personal level. However, workshop classes that focused on heavy machining and lifting, such as foundry and iron-smithy, drew sharp distinctions among genders; as a woman engineering student, I often felt both a sense of isolation and of excessive scrutiny in classes.

A general perception of a gender bias against women in "hard" engineering fields in the prevailing culture (outside of academia) was substantiated by my senior year internship experience in a local chemical plant. My entire internship was spent in a library with no assigned task, and, during my only tour of the factory, the engineer – seemingly proudly - described chemical pollution control devices that were turned on only at times of an external environmental inspection. The strong validation of women's knowledge pertaining to nature, health, engineering, and the environment that I had seen in my childhood years contrasted sharply with the instances of male dominance and apparent cynicism in engineering that I observed during my college years. This led to a shift in my career choice from chemical engineering to graduate study in environmental engineering in the US.

GRADUATE STUDIES IN ENVIRONMENTAL ENGINEERING IN THE US

As an environmental engineering graduate student in the United States, I was exposed to a world of freedom and choices that contrasted sharply with the prescriptive undergraduate engineering programs I had experienced. I was awarded a research assistantship for graduate study in the Civil Engineering Department at Carnegie Mellon University (CMU). With the Department of Engineering and Public Policy closely associated with Civil Engineering, I took a wide range of classes. The graduate classes in environmental engineering already covered a broad range of topics, from micro-scale water chemistry, macro-scale air pollution, climate change, and environmental systems modeling to environmental management.

I also took classes in ethnography, policy, and ethics that rounded out my graduate training and truly broadened my mind. I elected to take these classes to gain new experiences[9] in this new country. My fieldwork project for the class Ethnographic Field Methods began as an exploration of attitudes toward the environment in the steel mill town of Homestead, PA, and expanded to allow me to contrast attitudes toward poverty in the US and India. I took the courses in policy and ethics (I audited these) because these were offered in a very unique interdisciplinary Ph.D. program in Engineering and Public Policy (EPP) offered at CMU. Several faculty in Civil Engineering also taught in Engineering and Public Policy and there was an active exchange of students, faculty, and thinking between the two programs. For a while, I considered getting a Ph.D. in EPP, but I decided against it. No doubt the courses I took at EPP have had a lasting impact on me as I still have my

---

[9]I also took classes in Swimming and Film Appreciation to learn things that I had not experienced before. These are probably a result on my own interest in trying out different things that I had not been exposed to in India, as well as the sense of intellectual freedom and curiosity that I experienced at Carnegie Mellon.

class notes and textbooks, and some of my favorite readings come from some of these classes, which covered a range of topics from decision theory, to statistics, to ethics and the history of the role of science in US public policy.

My MS and Ph.D. research topics were equally diverse. My MS research focused on stochastic modeling of a large aquifer in the southwestern US. I immensely enjoyed the rigor of the mathematical theory and its computer aided implementation, and learned lasting lessons about mathematical modeling and data analysis that I have used again and again in all of my future projects. However, I came to realize that all the assumptions underlying our very sophisticated stochastic model, while very valuable as a thought experiment, could never be verified for "truth" without collection of very large, indeed impossible, amounts of data. For my Ph.D., I sought more hands-on laboratory work and direct data collection for hypothesis testing. These internal deliberations, which contrasted my math brain with hands-on work, resulted in further self-knowledge of my own research values.

I was fortunate that my advisors respected my personal search for research values and empowered me to seek and change focus for my Ph.D. dissertation. My dissertation focused on very micro-scale examination of mass transfer and bioavailability of pollutants to microbes during bioremediation of coal tar. I enjoyed experimental work in the lab, and there was a strong real-world application of this project in the field that attracted me. Further, I thought at that time that lab work had a higher level of rigor than more "hand-wavy" modeling[10], for which I had not yet developed the core confidence.

In the end, I graduated from Carnegie Mellon understanding the fun, frustrations, and limitations of both experimental and model-based research. For me, this combination provided the ideal research experience. The freedom to pursue my (research) values in the US had opened up a new window across my two cultures – a largely closed hierarchical Asian culture where young people rarely challenged pre-assigned paths set by elders (or teachers) versus a more open US culture where I had the freedom to express who I was and change the trajectory of my activities to be consistent with my core. By exemplifying academic freedom and mutual respect, my MS advisor demonstrated how educators can empower students to *practice* personal integrity and grow as individuals. Looking back, both curriculum and research experiences in graduate school shaped my future trajectory as an engineer and as an educator, but I would not realize this for more than ten years. (Clearly, self-knowledge seems to come to me very slowly!) I did realize, as soon as I became a faculty member, that I hoped to engage my own students with the same respect for their aspirations and support their discovery of their own paths as my advisor had with me in graduate school.

---

[10]To this day, I advise my graduate students that while laboratory work is frustrating, in a well-designed experiment the results that one gets are what they are and speak for themselves. Model development, on the other hand, needs graduate students to know the nuances of their models, the limitations, the boundaries and to know that some important factors may be unknown or too complex yet to grasp. As a Ph.D. student, I wanted the rigor of experimental work without having to "hand-wave" away things that could be important, but that were too complex to model and more importantly to verify against "reality."

LEARNING TO INTEGRATE ACROSS DISCIPLINES AND CULTURES

Moving from Pittsburgh to Colorado in 1994 - first, on a research faculty position and then as tenure-track faculty - was a smooth transition professionally, although more disruptive personally. I was surprised I could feel the difference between Pittsburgh and Colorado almost as much as I had felt the change from India to the US. Colorado just seems to have so much more of *Mother Nature* accessible to her people. Lots of sunshine, blue skies, mountains, streams, and meadows – all within a stone's throw from anywhere I lived. These provided great environments for contemplation. Socially, Coloradans seemed warmer – smiles when passing by anyone on the street, barefoot kids walking into each others' homes to play (reminding me so much of India), strangers passing by on walks would comment on how cute my baby son (and, later, my dog) were, and my neighborhood helped keep it all together. I doubt I would have coped with being a single mom and a young professor without the support of so many friends and neighbors who would coordinate car pools for our kids' summer camps, basketball games, etc. with my various deadlines for NSF and EPA proposals!

Perhaps because of the smaller number of Indians in Colorado or because I had grown away from the Indian (IIT-ian) graduate student subculture, I started developing a much clearer link with contemporary American culture (at least, as practiced in suburban Front Range Colorado), than I had during my years in Pittsburgh. I appreciated many things "American" from the life stories of my friends who had grown up in Colorado or other US states, and simply from everyday life. On my road trip from Pittsburgh to Colorado, I also came into contact, for the first time, with the Native American experiences in the West. I could now see both sides of the Columbus Day debates ongoing in Boulder and Denver. Perhaps I was getting assimilated, and at the same time, I had the opportunity to delve deeply into Indian (Hindu) philosophy, being responsible as a volunteer teacher for transmitting this heritage to our first generation Indian-American children in the US. Through all these social interactions and obligations, I was learning more than ever, unconsciously crossing many cultural lines and gaining confidence in stepping out of whatever boxes I was perceived to be in.

Professionally, I found myself once again surrounded by supportive colleagues and the most exemplary people. In my tenure-track position at the University of Colorado Denver, I discovered the joy and the hard slog of developing an independent engineering research program from scratch. I initially focused on small-scale bioreactor research and later expanded to study pollutant transfer in full-scale trees, in the then-emerging field of phytoremediation. My research group had interesting fieldwork projects. For example, one in which we visited a mom-and-pop gas station, whose wonderful owners were at risk of losing their retirement savings to remediation of a gasoline-spill from their station. We hoped our proposed solution (planting poplar trees for clean-up) would help their situation.

My work was getting more integrative and directly linked to people and economics. I was also fortunate to collaborate with another colleague and my MS advisor (all of us connected through Carnegie Mellon) on a book titled *Integrated Environmental Modeling* in which we present a unified framework for pollutant transport in air, waste, and soil, and impacts on biota (plants and humans). Writing the book was a five-year long, seemingly never-ending adventure, sometimes a chore, but in the end, a source of pride in our work and in our collaboration. Likewise, at UC Denver, the small size of our department seemed to also support more collaborations, including across disciplines of environmental, transportation, and structural engineering. Having been set free to seek my own path over all these years, I was having a lot of fun thinking up interesting projects that cut across disciplinary boundaries.

A LEAP TO SUSTAINABILITY SERVICE PROJECTS IN THE "DEVELOPING" WORLD

The leap from engineering with living plants and trees to "green engineering" for sustainable community development came after my tenure was awarded in 2003. A chance phone call started me on larger-scale "green engineering" projects focused on engineering for the developing world through a local non-governmental organization. Now, as a middle-aged professor and sustainability engineer, my students and I were tasked with providing clean water and renewable energy to tribal villagers in India.

As always, I was ready to try something new. The real-world application of engineering in developing community projects, and the potential to have a positive impact on under-served communities, resonated with me deeply. Growing up in India, I had experienced the positive side of Indian society in terms of social cohesion and a very profound (and holistic) Hindu culture, but I had also witnessed extreme poverty, cruelty to animals, and the sheer stress of vast population growth on tenuous natural resources. A chance to help in developing energy and water resources for community use, in a manner that preserved nature, connected supply-side technologies, the user sphere, and the environment, resonated with me intellectually, philosophically, and professionally. More profoundly, the human impact of these projects tugged at my heart and soul, something I caution aspiring young sustainability engineers about because, as I describe next, good intentions do not always correlate with rigor nor with desired outcomes.

Our site visit to the village was transformative. The lives of the villagers so closely connected with nature, their strong social networks, and inspiring participatory processes forced upon me, once again, questions of personal integrity and values. What does our US team of engineers really offer to a remote village in another country in an unknown social and political landscape? How can we engineer systems that will not disrupt local cultures? Who decides what is appropriate?

I remain conflicted on these questions. On the one hand, I am concerned too many humanitarian organizations present poverty in the "developing world" in a patriarchal, condescending manner of "us" helping "them," as a problem to be solved by the outsider. Our experiences in the tribal village in India made it clear the village was rich in natural resources, social networks, and indigenous holistic knowledge. They practiced rain-fed agriculture and, in good years, produced sufficient food for a whole year. The forests were sustainably logged. Tribal customs and communal activities were followed to support more arduous tasks such as home construction. And all decisions were made by complete consensus in community meetings attended by men and women – both equally vocal in debating the issues at hand.

But the villagers were clearly facing some physical hardships and political struggles, and they sought external assistance. While we did provide technical assistance by building a wind generator with the village and our NGO contact, our contribution felt incomplete to me. We had no direct understanding of how we were impacting the lives of the villagers, as we were removed from them by two language translations. The real know-how to operate the wind generator remained not with the villagers at large but with the NGO representative, due to the complexity of the machine. We had little understanding of the dynamics between the villagers and the NGO. The NGO, in turn, built upon Gandhian ideal of self reliance, had its own world view of the "developed west" that conflicted with some core principles such as "intellectual property" upon which so much of our technological innovations are based. In the long run, politics and social change seemed to be more important than mere technology. Such change can only come from within.

During the same period as our sustainable development project in India, I also took on a technical advisory position for one of the US organizations that builds and connects US university engineering teams with community projects, often in the developing world. I had the opportunity to review the workings of the organization, the mechanics of student projects, the potential politics of the donors and benefactors in the project, and the existing procedures for evaluating project process and project outcomes. From the perspective of projects, I was concerned that at that time participatory processes were not emphasized in training project teams. Sustainability often meant fiscal sustainability – not environmental or socio-cultural sustainability. A student club may be hard-pressed to provide quality assurance for the team going abroad. There is little opportunity to determine what philosophies and values are practiced by the team in its interaction with communities in another culture. Lastly, the sticky issues of donors and benefactors arose. I believe that I and many others after me made significant contributions to improve the operational procedures for such organizations, improving the mechanics of project implementation significantly.

But the overall concept of US student-faculty teams partnering with governments and NGO's in the developing world continues to raise three important structural questions in my mind:

1. *Who sets the agenda for the project?* Often it is local NGOs and/or their associated donor organizations that become international partners to US teams. Are multi-national corporations appropriate partners in development projects, and what is their agenda? Are church organizations with a proseletyzing mission in foreign countries appropriate partners? If the partner is an indigenous grass-roots movement, what is their specific agenda? Assuming democratic participation is valued, how does one ensure the partner represents the community? Is the US team knowledgeable and equipped to deal with gender, race, caste, religion, and other local political realities on the ground?

2. *Whose knowledge counts in solution formulation?* Is indigenous knowledge really represented in the projects? How can the US team be trained in participatory processes that facilitate community ownership and integration of local knowledge with any proposed engineering solution? How can

the US team be trained in looking at engineering solutions via multi-cultural lenses? Are there ways to future-cast and anticipate unintended consequences of technology choices?

3. *Who really benefits from the project and how are these benefits measured?* Given the various stakeholders involved in international sustainable development projects (businesses, US universities, the "beneficiary" community, its benefactor, or the local NGO), the question of who benefits, and how, is important to answer. However, understanding and measuring these benefits and costs can be quite challenging. Agenda-setting depends on who is at the table. For example, Figure 7.1 shows a community meeting in which the tribal chiefs (all men) represented their communities in meetings with UC Denver team members and proclaimed renewable energy to be their community's primary need. As we learned, this was because word had spread about a functioning wind generator we had built with and for a neighboring village; the tribal chiefs assumed that this was all that our team could do and wanted the same "gizmo" for their own village. However, based on best practices recommended for participatory rural appraisal, the UC Denver team sought a separate meeting with women only. The needs expressed by the women were markedly different from those articulated by the men, as water shortage in the dry season was expressed by most women as their primary challenge.

QUESTIONING THE SERVICE MODEL FOR INTERNATIONAL SUSTAINABLE DEVELOPMENT

I started reading up, independently, on the challenges associated with implementing service projects for sustainable community development in international settings. I expected that organizations like the World Bank and the UN would have encountered the same challenges decades ago and would be experimenting with alternate development paradigms by now. I found several articles and papers on these topics, both from the perspective of program outcomes assessment and the effectiveness of NGOs in international development projects. These articles echoed many of my own concerns outlined above, plus many more. Consider some examples.

- A recent cross-country macroeconomic study on the effectiveness of international aid summarizes existing literature on the topic and proposes a new analytical technique to more accurately discern the effectiveness of aid amongst confounding factors. It finds "little robust evidence of a positive (or negative) relationship between aid inflows into a country and its economic growth...no evidence that aid works better in better policy or geographical environments, or that certain forms of aid work better than others."[11] The authors in a separate paper explore why the beneficial effects of aid may be offset by other factors.

- Several papers discuss the rise of NGO's as key players in the alternate international development paradigm that arose in the 1980s. NGO's were viewed favorably over state actors due to the perceptions that they would be less "top-down," would engage in participatory processes,

[11] Raghuram and Subramanian, "Aid and Growth," 2005, 1.

Hamlet Leaders

Figure 7.1: Community meeting with tribal chiefs (top) and community meeting with women (bottom).

and provide for more flexibility and creativity in developing effective pro-poor policies. After more than a decade, however, NGO's are found to be facing many challenges that parallel those faced when US teams of globalizing engineers set out on international sustainable development projects. For example, Collinwood (2006) questions the legitimacy of transnational organizations because they are not held accountable to citizens in participatory process or by elections.[12] We found the same questions arose about the various NGO's who represented the projects our US team worked on – we had no way of knowing if and how much they represented the communities. Howell[13] notes that the effectiveness of NGO's is limited by five factors: an illusion of plurality and social inclusiveness, lack of experience in identifying all relevant pro-poor social actors on the ground, donor dependence and agenda-setting, elitism of the international partners chosen, and trying to apply a universal development template to local contexts where these may not apply. Howell specifically addresses the lack of political savvy in international project implementation: "Most donor agencies invest few resources in preparing their international field-staff for field assignments, let alone in understanding the intricacies of different political systems, local political histories and cultures. As a result, most field-staff grapple in the dark, trying to make sense of the plethora of civil society organizations they encounter." Agenda-setting by NGO's and their Donors (Project Benefactors) is addressed by Najam (1996) who argues that due to the large amounts of money made available to NGO's, "the intellectual undertaking of NGO enquiry remains predominantly donor driven."[14] Zaidi cites arguments made in the literature that, "the fads that are important for donors become translated into the policy agenda for NGOs in under-developed countries, as has been the case with environmental issues, a current favorite of the donor-driven developmental set."[15] These factors inhibit local participatory process and integration of local knowledge. The elitism of international project partners is described by Howell, who notes that partner choice is significantly influenced by "social ease with foreigners, command of donor discourse and physical proximity to donor offices." Such elitism then has the following effect:

> Donors may inadvertently reinforce social inequalities, contributing minimally to the strengthening of organization by the poor and the poor's capacity to articulate their concerns. Moreover, . . .donors may unavoidably become locked in particular elitist and clientelistic networks. The issue here is that such clientelistic ties steer donors towards particular sub-groups of the poor embedded in patronage networks from village level upwards. This then hinders the development of a generalized strategy to address poverty...[16]

[12] Collingwood, "Non-Governmental Organizations, Power and Legitimacy in International Society," 2006.
[13] Howell, "In Their Own Image," 2002, 125.
[14] Najam, A. "NGO Accountability," 1996, 342.
[15] Zaidi, "NGO Failure and the Need to Bring Back the State," 1999, 264.
[16] Howell, "In Their Own Image," 2002, 127.

Given these various conflicts of interest, I was heartened to find that NGO's working in the arena of global health are openly acknowledging these challenges and are proposing an NGO Code of Ethics for Health Systems Strengthening.[17] The code is motivated by the observations that, "[o]ften donors pressure NGOs to produce short-term gains in a limited population, creating conflict with the longer-term and more difficult task of building strong, high-quality national health care systems . . .," and that NGO's face the more difficult challenge "to 'do no harm': that is, limiting their own potentially negative effects on the public health system. The process requires NGOs to honestly assess their own practices and their unintended consequences and to recognize that working though the public system often takes longer and requires the NGO to share decision-making power."[18] These articles clearly articulate the challenges of international sustainable development work though a donor- or service-based NGO model. Reading them made me feel depressed to know that the large body of knowledge in the development area was not readily accessible or available to engineers either in their curricula or their international practice. Equally worrisome was why engineers working in international development did not appear to be seeking out this literature on their own accord. Was this due to the silo-nature of our curricula and professions and the attendant fear of being a neophyte in another discipline? I await the time when international engineering service organizations adopt a uniform code of ethics as have the health-focused organizations. Until that time, however, my own experiences with international projects and my learning from the above readings made me feel increasingly reluctant about embarking on many more international service projects.

As many have observed, transformational change in international communities comes from the work of citizen groups and NGOs on the ground. External agents (including US engineering teams) may offer funds or technical contributions, but they could also have their own unintended consequences and foster an overly skewed donor-recipient relationship, as numerous development case studies have documented. I personally felt I needed to have a better understanding of community participatory processes, the politics of agenda-setting, and partner-dynamics before pursuing further international *service* work. I highlight the word *service* as I find trans-country business partnerships, research partnerships, or education partnerships likely do not have quite the imbalance implied in service models.

DEVELOPING MULTI-DISCIPLINARY COMMUNITY-BASED SUSTAINABILITY TRAINING FOR ENGINEERS

I felt that before our UC Denver team could go out to "do good," which is a heartfelt instinct to be valued in all of us, we would need to have the patience to prepare well and learn across disciplines. I sought to develop a multi-disciplined curriculum for our burgeoning sustainability engineering program that connected engineers with four important concepts:

---

[17]Pfeiffer et al., "Strengthening Health Systems in Poor Countries," 2008.
[18]The NGO Code of Conduct for Health Systems Strengthening. http://ngocodeofconduct.org/

1. The politics of agenda-setting in international infrastructure projects and its relevance in international partner selection;

2. Participatory processes for self-defining community goals, and tools for measuring the impact of technologies on societal goals;

3. The solution-space that recognizes the (unintended) consequences of technology – environmental, economic, social, and cultural; and

4. Practical hands-on experience exemplifying the role of the engineer in societal decision-making.

Fortunately, despite the intensity of the service projects I was engaged in, I had retained my focus on green engineering research projects in the US. In 2003, I was the Principal Investigator (PI) on a large project awarded by the US Department of Education's Graduate Assistance in Areas of National Need Program (GAANN), in which I and three faculty colleagues (Rens, Janson, and Johnson) advised five Ph.D. students in the inter-disciplinary thematic topic of sustainable urban infrastructure. I found that, given my own city-upbringing, I had a better understanding of the infrastructures in urban Denver and their social ramifications than those in remote rural areas in India and other parts of the developing world. I also found myself relieved to be back in the realm of applied research in sustainable development, as the dynamics of donors, benefactors, beneficiaries, and their varying agendas in international service projects had caused me to lose much sleep over their ethical ramifications. The multi-discipline curriculum developed during the course of the GAANN grant included two courses that have since morphed: Field Methods for Rural Sustainable Development, and Environmental Life Cycle Assessment. Both of these courses are described in more detail shortly.

As my thinking about sustainability deepened, I also found myself distinguishing between global sustainable development and international sustainable development. In global sustainable development, we examine local actions in the US from the viewpoint of global impacts such as climate change, energy security, etc. The globally-aware engineer recognizes that local actions have global impact on people and the planet, and vice versa. In contrast, the international development engineer implements infrastructure projects internationally, but these may or not have an explicit global impact focus. In either case, sustainability criteria – the three E's of sustainability: economics, environment, and equity[19] - require that the sustainability engineer recognizes the links among physical infrastructures, natural environment, and social actors.

Our 2003 GAANN grant provided funds to build a curriculum around the idea of urban infrastructures with a global sustainability emphasis. In 2005, based on our project experiences in India, our team won EPA's National P3 Student Design Competition for developing a renewable energy systems design for remote rural locations in South Asia.[20] P3 refers to another variant of the three E's – People, Prosperity, and the Planet. Both of these grants provided the impetus for

[19] Campbell, "Green Cities, Growing Cities, Just Cities?" 1996.
[20] Ramaswami et al., "Sustainable Energy Systems Design for a Tribal Village in India," 2005.

developing multi-disciplinary coursework pertinent to sustainability engineering with a global and/or international focus.

I had many conversations with my colleague, Mike Tang, who teaches history of technology at UC Denver. He introduced me to the works of Arnold Pacey, David Hess, Jerry Mander, and others in the science and technology studies (STS) field. I recalled other readings from my coursework in ethics and in ethnography a decade earlier at Carnegie Mellon, including the works of Carolyn Merchant and Vandana Shiva. Our project in India had provided us with a real-world case study that highlighted the technological, social, cultural, and language challenges in international sustainable development projects. I used all these materials to develop a course on Fieldwork for International Sustainable Development in 2005. Table 7.1 summarizes the principles of green engineering that informed my course development.

Based on these principles, I developed two courses for the GAANN program. Field Methods for Sustainable Development discusses the many definitions of sustainable development and notes that sustainability must necessarily be locally defined. Articles by writers such as Vandana Shiva, Arundhati Roy, and Jerry Mander describe some of the challenges that arise when local voices and cultures are ignored in planning infrastructure projects – largely focused on rural projects. Techniques for fostering community participation in various models – Participatory Rural Appraisal, Participatory Action Research, and Community-Based Participatory Research – are presented to the class and exemplified in terms of project experiences (e.g., Figure 7.1). Students then learn to do a resource assessment in rural communities. This may include measuring wind speeds to assess wind energy potential and then hands-on construction of engineered solutions and evaluation of their "greenness" according to the Principles of Green Engineering[21] (See Table 7.1). Long-term sustainable project management by local entrepreneurs is envisioned and is discussed in the context of micro-enterprise through guest lectures and case study readings. The course was developed for MS students in Engineering and Environmental Science who seek to practice sustainable development projects in the field. The primary focus of this course is to connect technologies with people.

A second course on Environmental Life Cycle Assessment (LCA) teaches students structured techniques that have now been in development for two decades to quantify the environmental impacts of various engineered products and processes. LCA makes transparent the environmental impact of all engineered products and processes from "cradle-to-grave," i.e., across the supply chain from resource extraction to product manufacturing, product use, and disposal or recycling (end-of-life). LCA and its companion tool from the field of industrial ecology, material flow analysis (MFA), provide the rigor that I sought in quantifying environmental sustainability instead of merely talking about "going green." Indeed, MFA and LCA analyses have often revealed many of the complexities of our production systems, for example, highlighting the unintended negative impacts of biofuels production through analysis of processes occurring upstream and downstream from biorefineries. Although incomplete, LCA and MFA tools connect technologies with nature in quantifiable ways.

---

[21] Anastas and Zimmerman, "Design Through the 12 Principles of Green Engineering," 2003.

Table 7.1: Principles of Green Engineering.

US Envrironmental Protection Agency[22]	Anastas and Zimmerman, 2003[23]
1. Engineer processes and products holistically, use systems analysis, and integrate environmental impact assessment tools.	Principle 1: Designers need to strive to ensure that all material and energy inputs and outputs are as inherently nonhazardous as possible.
2. Conserve and improve natural ecosystems while protecting human health and well-being.	Principle 2: It is better to prevent waste than to treat or clean up waste after it is formed.
3. Use life-cycle thinking in all engineering activities.	Principle 3: Separation and purification operations should be designed to minimize energy consumption and materials use.
4. Ensure that all material and energy inputs and outputs are as inherently safe and benign as possible.	Principle 4: Products, processes, and systems should be designed to maximize mass, energy, space, and time efficiency.
5. Minimize depletion of natural resources.	Principle 5: Products, processes, and systems should be 'output pulled' rather than 'input pushed' through the use of energy and materials.
6. Strive to prevent waste.	Principle 6: Embedded entropy and complexity must be viewed as an investment when making design choices on recycle, reuse, or beneficial disposition.
*7. Develop and apply engineering solutions, while being cognizant of local geography, aspirations, and cultures.	Principle 7: Targeted durability, not immortality, should be a design goal.
	Principle 8: Design for unnecessary capacity or capability (e.g., 'one size fits all') solutions should be considered a design flaw.
8. Create engineering solutions beyond current or dominant technologies; improve, innovate, and invent (technologies) to achieve sustainability.	Principle 9: Material diversity in multicomponent products should be minimized to promote disassembly and value retention.
	Principle 10: Design of products, processes, and systems must include integration and interconnectivity with available energy and materials flows.
9. Actively engage communities and stakeholders in development of engineering solutions.	Principle 11: Products, processes, and systems should be designed for performance in a commercial 'afterlife.'
	Principle 12: Material and energy inputs should be renewable rather than depleting.

[22]US Envrironmental Protection Agency, "What is Green Engineering."2010

[23]Anastas and Zimmerman, "Design Through the 12 Principles of Green Engineering."2003

In addition to courses developed in the GAANN Program, our experiences in our EPA project implemented in villages in India also raised another issue of the dynamic between "developed" and "under-served" or "indigenous" societies. Should we not be learning about sustainable lifestyles *from* such developing world knowledge instead of blindly transporting technocratic systems to them? Two colleagues, Jim Mihelcic and Julie Zimmerman, and I researched the first question and published two papers on knowledge transfer *from* the developing to the developed world. Jim initiated this wonderful project after we found ourselves meeting at conferences and lamenting the one-way transfer of knowledge, from the West to developing nations, often featured in student projects. We decided to use case studies rather than theoretical exhortations on co-learning with our communities in the developing world to illustrate that knowledge about sustainability practice has indeed flowed in from developing and indigenous communities to the developed world. We hoped that by illustrating this seemingly radical idea to students, they would naturally come to view developing communities as storehouses of useful, holistic information explicitly relevant to science, technology, economics, and governance, rather than passive recipients of western aid.

In researching these case studies, I was surprised to find out that the first biogas plant was developed in a leper colony in Mumbai, India, and the technology was then transferred by the British to wastewater treatment plants in England.[24] In addition to technology innovations, we were also interested in showing the practical value of social capital[25] in sustainability, beyond the fuzzy warmth of the community that most US teams notice immediately when they first interact with developing world communities. We wanted to show that these social networks lead to tangible economic and governance outcomes. In this context, I found myself fascinated by the Bali Rice-Farming case study, Elinor Ostrom's Eight Design Principles for cooperative commons governance, and the Grammen Bank's peer-lending networks – all three of which demonstrate, in different ways, how social capital contributes to sustainability in many communities worldwide.[26] The issue of spatial scale of governance and of environmental impact of infrastructures emerged. The case study research experience with Jim and Julie provided another leap forward in integrating concepts both across scale and disciplines.

Meanwhile, at UC Denver, our five GAANN students were doing wonderful research in life cycle assessment. LCA is an important tool that enables us to trace the environmental impact of products beyond the spatial and temporal realm of their end-use. For example, the environmental and ecosystem impacts of driving an automobile are traced from resource extraction (oil extraction for gasoline, iron ore extraction for the car parts, aluminum extraction for the frame, etc.) to manufacturing (metal forming, auto manufacturing, oil refining), operations (tail pipe emissions from

[24] See Mihelcic et al., "Integrating Developed and Developing World Knowledge into Global Discussions and Strategies for Sustainability," 2007.

[25] Social capital represents the benefits experienced from social connections and networks, similar to that provided by using economic capital (money) and natural capital (natural resources). Indeed, Viederman defines sustainability development as that which conserves five types of capital: natural, human, human-made, social, and cultural. See Viederman, "Sustainability's Five Capitals and Three Pillars," 1996.

[26] See Ramaswami et al., "Integrating Developed and Developing World Knowledge into Global Discussions and Strategies for Sustainability," 2007.

driving the car), to the grave (land-filling or recycling of waste materials from the car). LCA is a holistic way of connecting human use of commodities back to our overall impact on nature. The question that often arises is what impacts on nature do we care about? Do we care more about local air pollution, global climate change, loss of biodiversity, release of hazardous waste, loss of agricultural land, consumption of water supplies, depletion of fossil fuels, or any other of a myriad of impacts we have on nature and thence on ourselves?

One of my Ph.D. students, Mike Whitaker, was researching LCA of bus transport infrastructure in India. As part of his fieldwork, we developed and implemented a survey trying to identify environmental priorities for people in Chennai, India – we realized that traditional LCA's were developed largely based on western environmental priorities. This work was received very well in India, although getting the technical manufacturing data to complete the research was a challenge. Meanwhile, in Denver, Ph.D. student Mark Reiner was examining the LCA of concrete use in urban infrastructure in US cities. These case studies improved the Life Cycle Assessment course (described previously) by providing contrasts in data and methods across two countries.

Graduate advising in the realm of sustainable development was very rewarding for me. Indeed, in contrasting project experiences working with undergraduate classes in our EPA P3 project with graduate research advising in our GAANN program, I found graduate level work allowed for more time and more student maturity in picking up nuances of sustainable development. For example, we tried integrating our village wind generator project with a few senior design courses (two in Electrical Engineering at UC Denver and one senior design Mechanical Engineering class at Colorado State University). The scope of our project in India was too large for effectively integrating participatory research at the undergraduate level; instead, longer-term the graduate students did this work. The participatory process itself took more than a year, while testing wind generator components even at the bench-scale was more than a two-year process, which did not even cover the necessary time (more than six months) needed to effectively test wind generators in field conditions. As the scope of the project changed based on needs expressed by the villagers (Village one wanted wind power while village two turned out to need water supply), tailoring each project to the actual participatory process was better accomplished by Ph.D. students who had a longer timeframe in the university. Here again the challenge of appropriately conducting participatory projects in international communities emerges – these processes must necessarily be slow to incorporate community input adequately. One cannot rush community processes to match the rather tight timeframes of undergraduate senior design classes. In addition to time needed for participatory community processes, I found both my graduate students and I needed many months to integrate and synthesize our learning from communities. Advising graduate students in this effort included many meetings in which we clarified and re-clarified the processes by which our UCD team would work with communities. Maintaining neutrality and, more importantly, a sense of "dispassionate interest" toward the choices made by communities while providing technical information about sustainable technologies was very difficult, even for graduate students, as it required self-examination and self-control. I am hazarding a guess that such maturity is likely even more difficult to develop at a younger age. In terms of learning opportunities, working with communities locally in Denver (described next) provided our GAANN graduate student team with the chance for more frequent meetings with communities than the few international trips we made to villages in India. More time at the graduate research level allowed for more technical rigor. In the next Section, I suggest that learning first at home in a research setting before going abroad perhaps can help avoid many of the pitfalls associated with international service learning projects in sustainable development. However, I am yet to have a definitive answer on this.

LEARNING FIRST AT HOME, BEFORE VENTURING ABROAD

Taking a lead from Ph.D. student Mark Reiner's interest in having a real-world impact for his LCA of urban concrete, one day I called the City and County of Denver to ask if we might collaborate with the City on Denver's sustainability efforts. Wonderfully, Beth Conover, then the Director for Sustainability at the City and County of Denver, readily agreed, and so began one of the most fruitful partnerships in which I have been engaged. This partnership continues today.

Beth Conover later took on the position of Director of Greenprint Denver, leading a group of roughly forty diverse citizens and stakeholders who were tasked by Mayor Hickenlooper to develop Denver's Climate Acton Plan. The goal of this Plan is to reduce Denver's per capita greenhouse gas emissions 10% below 1990 levels by the year 2012. My research group at UC Denver was contracted to provide technical assistance to Greenprint Denver both on computing a greenhouse gas inventory for the city and on policy options for carbon mitigation. At the research end, I quickly realized that cities need to incorporate into their planning such advanced tools as LCA in order to visualize their impact on nature, including beyond city boundaries. For example, while personal carbon calculators and national carbon emissions inventories often include the energy needed for food production and airline travel, these important activities are often missing from city-scale carbon inventories because food production and airports are frequently located outside the boundaries of many cities. We developed the research and published an important paper on improved methods for city-scale greenhouse gas inventories.[27] Eight other US cities found the work so interesting that they wanted to replicate the method. Through a collaboration with Chris Kennedy at the University of Toronto, the method has also been expanded to London, Geneva, Bangkok, Toronto, and New York.[28]

Even more exciting and transformational for me was learning and thinking about how best to communicate the technical information on energy use, conservation, efficiency, and renewables to the group of forty citizens and stakeholders in Greenprint Denver. I found watching Greenprint discussions on the pros and cons of various civic engagement strategies and policy options for sustainable water and energy in Denver enormously illuminating. When their final consensus recommendations were released for public comment,[29] I felt humbled and privileged to be part of this important community process. I realized once again that the rubber meets the road when technology and technical information are used by people both in policy-making and in everyday

[27] Ramaswami et al., "A Demand-Centered Hybrid Life Cycle Methodology for City-Scale Greenhouse Gas Inventories," 2008.
[28] Kennedy et al., "Greenhouse Gas Emissions from Global Cities," 2010, In review.
[29] Greenprint Denver, "City of Denver Climate Action Plan," 2007.

life. Energy conservation, energy efficiency, and adoption of renewable energy will not happen by technical virtuosity alone – the role of social actors is paramount.

In a new research project, we are currently working with the City and County of Broomfield to study the very process of communicating technical information to policy makers. Along with my colleagues Debbi Main and Chris Weible, who work in public engagement and public policy, respectively, we are studying the process of analytic deliberation in sustainability planning in cities in the US. Analytic deliberation[30] is a combination of analysis (often facilitated by "technical experts") and democratic deliberation to build consensus in developing plans or policies. Through surveys being administered to the City of Broomfield Sustainability Task Force before, after, and during their plan-making process, we will be evaluating knowledge gain, opinion change, and consensus-building as analytic deliberation proceeds. We expect that this research will provide valuable data on how technical knowledge is processed by decision-makers, and help answer if and how university researchers can facilitate analytic deliberation in communities. If found effective, the process developed by our UC Denver research team is expected to be replicated by the National Civic League as a model for communicating energy, carbon footprint, and sustainability planning strategies to communities in the US, using a participatory analytic-deliberative framework.

And so, since 2005 I have found myself very much at home in the US contemplating energy technologies with the Colorado community. Since 2005, my group has been requested to work with four more cities of diverse sizes and character in Colorado (Broomfield, Arvada, Aurora, and Central City) on sustainability planning. It is humbling to know that our work could become meaningful to more than one million people in Colorado.

FOR ME, THE ANSWER IS HERE

Rather than "teaching" other communities to be more sustainable, I now find myself learning first in my own. The scale of my work has shifted from small villages in developing countries to cities everywhere, beginning with my own city of Denver, CO. Our group at UC Denver is bringing technical information on greenhouse gas mitigation, energy efficiency, and water efficiency into the democratic deliberative processes occurring in city councils, community groups, and other public discourse forums in Colorado. Our rural work in India raised this question: How can engineers be agents of change in a global world while maintaining scientific and ethical rigor and practicing social-cultural sensitivity? I find the answer is here. We understand to some extent the culture and the dynamics of communities here in Colorado because we live in them – and we can bring our technical knowledge into social-political discourse right here. We satisfy the important pre-requisite of being co-located with our partner cities to engage in this important discourse. Yet, we have the space to step back into our journal papers to pursue the rigor of research that is so often lacking in the field of sustainability.

Likewise, the communities we seek to partner with are familiar to us, and yet each is uniquely different. We learn about each new community, including when we need to step back and leave

[30] Sachs, "Sustainable Development," 2004.

Figure 7.2: Analytic deliberations at a Greenprint Denver meeting (top), and community engagement activities for sustainability in Denver (Bikeshare Program in 2008). Courtesy Cindy Bosco, 2008.

the communities and councils to make their decisions. And, amazingly, while I imagine that I must appear an "outsider," I find myself being welcomed into cities and communities in Colorado as we grapple with issues of sustainable global water and energy use through local action. I am surprised by how comfortable and how much a part of the community I feel. Global is indeed local. As an engineer and as an educator, I am now discovering what it means to be *glocal* in my own community.

Trying to understand people (starting with self) and also understanding technology appear to be key elements in our effort to build a more sustainable world. Practicing both skills at the smaller local scale - even the home - may be the answer to the world's most pressing needs. Indeed, given the cultural and political challenges documented in international development work, it may be that students and faculty interested in developing the globally-aware engineer might want to train them first through fieldwork experiences in the US. We could be learning and practicing community engagement, participatory planning, and analytic deliberation here in our own communities of diverse cultures, before translating engineering design and implementation to other parts of the world.

In 2007, I proposed some of these ideas on linking engineering with social actors in a successful NSF IGERT grant proposal on the thematic topic of *Sustainable Urban Infrastructure*. Through this five-year $3.2M grant, we are recruiting about twenty-six Ph.D. students who will be researching the connection between technology and social actors in design, analysis, and dissemination of sustainable urban infrastructures of the future. The students will be conducting fieldwork in cities in the US. About half of the students have funds allocated to conduct parallel research in cities in India where they will be comparing future infrastructure trajectories, enabling technologies and transformative policies across the two continents. Cross-discipline curriculum in our IGERT program consists of three courses:

1. *Introduction to Sustainable Infrastructure:* An introductory course in which students learn about the principles, pathways, and measurement tools to rigorously measure progress toward sustainability using engineering, planning, policy, and human behavior change strategies. As an integrative example, the class addresses creating a more sustainable urban transportation system integrating vehicle-fuel advances (technology), urban and regional smart growth planning, carbon tax and other policy instruments, and, behavior change. Several faculty from the four disciplines (engineering, regional planning, public affairs, and behavioral sciences) provide lectures, integrated by me as the lead instructor. The class has been offered twice and has seen enrollments greater than twenty students each time, drawn from graduate students in the four disciplines.

2. *Defining and Measuring Sustainability*: A team-taught class in which we work on a real-world sustainability planning project with a city in Colorado. My colleague, Debbi Main, covers techniques for community engagement and participatory research. In this module, the students conduct focus group sessions and develop or administer surveys to understand sustainability aspirations of the community. The direct interactions between students and the community through focus groups has been cited as most transformative and educational for the students. In the measurement module, I provide the tools of LCA and MFA so that students can estimate a

baseline carbon emissions footprint for the city based on field data from water, transportation, and electric utility sectors.

3. *Planning and Policy for Sustainability:* The fieldwork data from Defining and Measuring Sustainability feeds into the next class on planning and policy for sustainability (in development). It is a team-taught class across these disciplines that exposes students to theories underlying policymaking, and, the implementation process in urban planning.

As I draft this chapter, I feel I have myself come "home" full circle, in many ways.

- Research versus Service: As much as international service projects are compelling, I find my intellectual home is in applied research in global sustainability where I can work with well-defined partners charged with implementing programs on the ground on a long-term and large-scale basis.

- Collaborations: My professional home seems to be grounded in collaborations. In writing this article, I realize the richness of my collaborations with many people over many years.

- Crossing Cultures: Having travelled and lived in many places, I must confess that I have no geographical home per se. I would call any community that I am integrated with my home.

- Crossing Disciplines: My disciplinary home is now aggressively multi-disciplinary. Perhaps, now, I can combine the formative experiences of my childhood by learning and integrating soft- and hard- paths for global sustainability, linking people, technologies, and nature.

Implementing, researching, and evaluating sustainable development in any setting is difficult as it tries to achieve that perfect balance between people, prosperity, and the planet. In my journey, I have found that all my learning over my lifetime – even from early childhood - has helped me in integrative holistic thinking. More recently, when exploring sustainability service projects in rural international settings (even in my own home country), I found that working with short contact times, and in unfamiliar, social-political environments can pose challenges. At this time, I am hopeful that *first* training students in community-partnered sustainability projects in the US, inculcating rigor through research in both the physical and social sciences, and *then* partnering with elected local governments in international cities may provide a balanced and scalable approach toward global sustainable development promoted by faculty-student research teams at US universities. However, only time and our experiences over the next four years will tell us if this hypothesis will be validated. If yes, the potential for large-scale global impact is tremendous – from impacting about 800,000 people here in Colorado today, to helping facilitate collaborative analytic-deliberative technological change for millions living in cities globally. If our program in Sustainable Infrastructure at UC Denver can provide a template, such an effort will involve large numbers of students and faculty transcending the divides between research and service (Why can't research projects be of service?), national versus international (Can we learn here before we go abroad?), expert versus user (Can universities engage in analytic deliberation?), and, between the physical and social sciences, connecting technology with people.

REFERENCES

Anastas, Paul T. and Julie B. Zimmerman. "Design Through the 12 Principles of Green Engineering." *Environmental Science and Technology* 37, no. 5 (2003): 94A – 101A. DOI: 10.1021/es032373g 164, 165

Campbell, Scott. "Green Cities, Growing Cities, Just Cities?" *Journal of the American Planning Association* 63, no. 3 (1996). 163

Collingwood, Vivien. "Non-Governmental Organizations, Power and Legitimacy in International Society." *Review of International Studies* 32 (2006): 439–454. DOI: 10.1017/S0260210506007108 161

Gleick, Peter H. "Global Freshwater Resources: Soft-Path Solutions for the 21st Century." *Science* 302, no. 5650 (2003): 1524 –1528. DOI: 10.1126/science.1089967 153

Greenprint Denver. "City of Denver Climate Action Plan." 2007. http://www.greenprintdenver.org/docs/DenverClimateActionPlan.pdf. 168

Howell, Jude. "In Their Own Image: Donor Assistance to Civil Society." *Lusotopie* 1 (2002): 117–130. 161

Jayaram, S., S.P. Thyagarajan, S. Sumathi, S. Manjula, S. Malathi, and N. Madanagopalan. "Efficacy of Phyllanthus Amarus Treatment in Acute Viral Hepatitis A, B and non A non B: An Open Clinical Trial." *Indian Journal of Virology* 13, no. 1 (1997): 59–64. 151

Kennedy, Chris et al. "Greenhouse Gas Emissions from Global Cities." In review at *Environmental Science & Technology*. DOI: 10.1021/es900213p 168

Lovins, Amory. "Energy Strategy: The Road Not Taken." *Foreign Affairs*, 1976. 153

Mihelcic, J.R. J.B. Zimmerman, and A. Ramaswami. "Integrating Developed and Developing World Knowledge into Global Discussions and Strategies for Sustainability. Part 1: Science and Technology." *Environmental Science & Technology* 41, no. 10 (2007): 3415–3421. DOI: 10.1021/es060303e 166

Najam, Adil. "NGO Accountability: A Conceptual Framework." Development Policy Review 14, no. 4 (1996): 339–354. DOI: 10.1111/j.1467-7679.1996.tb00112.x 161

Pacey, Arnold. *The Culture of Technology*. Cambridge: MIT Press, 1983. 152

Pfeiffer, J. et al. "Strengthening Health Systems in Poor Countries: A Code of Conduct for Non-governmental Organizations. *American Journal of Public Health* 98, no. 12 (2008): 2134–2140. DOI: 10.2105/AJPH.2007.125989 162

Raghuram, G. Rajan and Arvind Subramanian. "Aid and Growth: What Does the Cross-Country Evidence Really Show?"Cambridge, MA: National Bureau of Economic Research Working Paper No. 11513, August 2005, Revised February 2007. DOI: 10.1162/rest.90.4.643 159

Ramaswami, R., J.B. Zimmerman, and J.R. Mihelcic. "Integrating Developed and Developing World Knowledge into Global Discussions and Strategies for Sustainability. Part 2: Economics and Governance." *Environmental Science & Technology* 41, no. 10 (2007):3422–3430. DOI: 10.1021/es060303e 166

Ramaswami, A., T. Hillman, B. Janson, M. Reiner, and G. Thomas. "A Demand-Centered Hybrid Life Cycle Methodology for City-Scale Greenhouse Gas Inventories." *Environmental Science & Technology* 42, no. 17 (2008): 6456 – 6461. DOI: 10.1021/es702992q 168

Ramaswami, A. et al. "Sustainable Energy Systems Design for a Tribal Village in India." 2005. http://cfpub.epa.gov/ncer_abstracts/index.cfm/fuseaction/display.abstractDetail/abstract/7173/report/0. 163

Sachs, Jeffrey. "Sustainable Development." *Science* 304, no. 5671 (2004): 649. 169

The NGO Code of Conduct for Health Systems Strengthening. http://ngocodeofconduct.org/

US Environmental Protection Agency. What is Green Engineering. 2010. http://www.epa.gov/oppt/greenengineering/pubs/whats_ge.html. 165

Viederman, S. "Sustainability's Five Capitals and Three Pillars." In *Building Sustainable Societies: A Blueprint for a Post-Industrial World*, Edition 1, edited by Dennis C. Pirages, Chapter 3.M. E. Sharpe Inc, 1996. 166

Zaidi, Akbar. "NGO Failure and the Need to Bring Back the State." *Journal of International Development* 11, no. 2 (1999): 259–271.
DOI: 10.1002/(SICI)1099-1328(199903/04)11:2%3C259::AID-JID573%3E3.0.CO;2-N 161

CHAPTER 8

If You Don't Go, You Don't Know

Linda D. Phillips

There are simply not enough resources for 6.7 billion people to live the way the developed world lives. We see constraints with rising food prices due to unstable supply, rising oil prices, and numerous other issues driven by the reality that the world is not currently capable of providing a sufficient level of resources to the entire population, at least not without severe consequences.
—International Senior Design Student Learning Log

INTRODUCTION

The past decade has seen an awakening among young people who wish to reach out to humanity and make their world sustainable. In fact, the top four issues on the minds of the US millennial generation are education, poverty, environment, and health and disease.[1] They wish to work a high level of "green awareness" into their personal and professional lives, are fascinated with innovation and technology, and also want to eradicate poverty and hunger. Fortunately, for the engineering community, this vision for change in young people's desires can be aligned strategically with national and international goals that have also documented rapid social, political, economic, and environmental changes occurring in the world. The young people's new social consciousness has implications for engineering education, research, practice, and, importantly, the economic competitiveness of our nation and the world.

I have witnessed firsthand students' dedication to global values—it is they who pushed for the International Senior Design capstone course at Michigan Tech. It is students such as these who motivate me to continue opening eyes, hearts, and minds through the program, providing young people a venue for using their education to address global issues.

In this chapter, I trace the pathway that led me to develop International Senior Design at Michigan Technological University, which my husband and I directed. In late 2008, we transferred the program to the University of South Florida, where it is called International Capstone Design. However, since I focus on the life of the program at Michigan Tech in this chapter, I use the acronym ISD.

[1] Cone Inc. and AMP Insights, "Cone Millennial Cause Study," 2006.

Service learning is experiential education designed to serve the needs and interests of a community. ISD students work in developing world communities, often without modern technology, using their own creativity in addition to local wisdom to adapt to local practices. ISD addresses the global economy in which graduates will work, providing engineering experiences that can demonstrate to students that engineers' social responsibilities are fundamentally important and address real facets of sustainable engineering, including global and community awareness. This is achieved by presenting environmental, economic, and societal issues, as well as opportunities for intercultural and interdisciplinary interaction.

Although I developed the original course and structure, ISD evolved through my responses to various requests from undergraduate and graduate students across campus, project benefactors, local in-country hosts, faculty, and Michigan Tech's Department of Civil and Environmental Engineering. My own needs and requests also produced opportunities for key participants, situating them in unknown or novel territory that motivated them to embark on a journey of discovery, whether abroad or on campus. I have continued to respond to requests with openness, flexibility, and, at times, perseverance against traditional university mindsets. Many students and alumni report that they consider the class experiences, both its US and its abroad components, life-changing and more valuable than similar, more-traditional efforts on campus. These students and alumni have been emphatic about this, more so than I imagined. With that said, I designed ISD to be not just a capstone design course, but also a crash course on life for many students. That alone, however, will not satisfy ABET criteria or university requirements, as I have found.

I begin with my background and then describe the ISD program. To help readers understand both successes and struggles, I give examples of various requests and responses that shaped the program. Finally, I offer my reflections on the impact of the program, including what it does or does not do, as well as its growth.

TRANSITIONING FROM INDUSTRY TO ACADEMIA

As a young child, I had the passion to teach and wanted to follow in the footsteps of my parents as an educator. At my father's counsel and after gaining construction experience while building our own home, I chose an engineering path "to solve math problems on my own level." However, I never solved those mathematical design problems, as I was busy managing projects instead of designing them. With an MS degree in Construction Management, I spent over twenty years managing projects in the design and construction industry. I climbed to a vice president position, supervising project managers in the western half of the US.

Following my original passion, though, I continuously sought opportunities within industry to mentor, train, create workshops, and teach. I believe I was born to teach, and although my first career took me in another direction, learning to manage projects and people prepared me well for the opportunity to open students' eyes, minds, and hearts while meshing engineering, service, and sustainability.

After climbing the industry ladder of success, I grew discontent clearing and paving green spaces with unneeded retail space or coaching and managing my employees to do the same. I felt I was encouraging the expansion of an already overbuilt environment. How much can we fertilize the soil of our nation with commercialization before it becomes overgrown? I had an uneasy feeling about continuing on the path of industry, and I knew it was time to search for something else, although I did not know for what at the time. I longed for something more meaningful—something that could impact the world in a positive way.

The request came in the spring of 1997: "Come to Virginia Tech for a semester and teach the professional practice contracts and specifications course." "Oh, I can't," was my initial response. "I have a job, a husband, a house." But the desire to teach won, and I took a leave of absence.

Before leaving for Virginia, I found myself in tears, asking, "How can I teach students to 'develop' more unneeded facilities?" I would be teaching engineering students the practices I had learned in industry: designing and building more of what we do not need here in the US. Without other immediate options, I based my teaching on my experience in US construction projects and typical business practices. In classroom discussions, however, I found myself interjecting lessons from my last (and toughest) industry project: a retail center on a developing Caribbean island. There I had learned that Haitians valued having a job each day more than the pay itself. Although they were paid per piece, for them it was more honorable to work slowly and make the job "last" than hurry the installation to get final pay. I also learned the importance of alternative designs for rainwater harvesting and sewage treatment and the importance of designing pipe sizes to be "nested," minimizing shipping costs to the island. And sadly, I learned some of the consequences of a litigious and greedy US society. The project/facility owner sued, taking advantage of the situation to earn a fast buck. Happily, I had stirred student interest.

During the same time period, I participated in two volunteer construction projects with my mother in South America. My mom, a former primary-through-college-level educator, had done construction on international mission trips since 1989, after my father passed away. On our trips, I challenged the traditional mindset that the US women should straighten nails while the US men "constructed." After one day of straightening nails, I climbed the ladder, swung the hammer, learned unconventional forming layout, and helped with mix designs for hand-mixed concrete. Through this work, I met a missionary in Bolivia who, upon learning I was a civil engineer, requested that I help him develop a site plan for a proposed school, church, and community center. At the time, ISD was an intangible dream, so I never imagined this request would turn out to be the exigence for the first ISD class project.

DEVELOPING ISD

When I returned to industry from Virginia Tech, I was greeted with a pink slip (luckily, in retrospect), and in the fall of 1998, I was asked to teach the civil and environmental engineering professional practice course at Michigan Tech. Based on my volunteer experiences, I reformatted class assignments using detailed photos of developing-world construction materials and techniques, unintentionally planting the seeds of ISD.

During each subsequent professional practice course, the same student request continuously arose: "We've written the specifications, so why can't we go build our projects?" So, for two years, I researched options within the university, making proposals to the study abroad office. I lobbied the civil and environmental engineering department, the dean, and anyone who would listen—all to no avail. All anyone could see was a "nontraditional," non-tenure-track employee on a one-year contract proposing a service-learning program that posed risk and offered little educational value in terms of traditional course credit.

In May 2000, a usually timid student entered my office and insisted that "we go build this." Not wanting to reveal that my teaching contract had not been renewed, I told her it was not feasible without students. "How many?" she asked. And without thinking, I quickly responded, "Fourteen!" She left telling me she would have them the next day. I ignored her, assuming it was impossible for someone so bashful to find even three people to agree to go to Bolivia. She returned the next morning with seventeen signatures, and, again in a panic, I sent her to the student advisor—anything to divert her, as my appointment was days from expiring.

At the same time, the department head was facing a new issue: the new Engineering Criteria 2000 ABET requirement. Our current curriculum needed capstone design courses, yet there was a shortage of professors who cared to teach them. So the department head was now interested in the students' pleas. I quickly redrew proposals and budgets. He approved them and offered me a single-course teaching contract. By July 2000, twelve student deposits were in hand and we had an engineering problem to solve: create site plans and designs for the new Bolivian school site, which flooded with every rainstorm.

Since I needed a project quickly, I chose that particular location in Santa Cruz, Bolivia, in conjunction with the missionaries. I was familiar with the camp-style accommodations and the proposed design project, and there was a construction project nearby. Together we completed logistics, and the ISD course took-off ahead of the times: two years before the Millennium Development Goals and Engineers Without Borders-USA were established and five years before the Michigan Tech Engineers Without Borders chapter was formed.

Since that first project in 2000, the capstone ISD has provided a service-learning design experience situated in the developing world. Students explore the technical, economic, and social implications of engineering design and construction through projects such as: improving water supply, water resources, and storm water management; site master planning; site reclamation; solid waste management; and wastewater treatment. These projects have been located in communities in Bolivia and the Dominican Republic.

EXPANDING ISD: FROM THREE TO SIX CREDITS

Originally, I designed the ISD course as a summer-semester, three-credit, major design experience that could also fulfill technical elective requirements. It grew to six credits with the option of an undergraduate certificate developed in 2007.[2] Since I was funded by summer tuition, both three-credit courses were scheduled during the summer.

In the first portion of ISD, students spend at least two weeks in-country, partly working on a construction site to learn the local tools, materials, and techniques. Each day, students do construction labor alongside local workers, helping administer and manage the daily construction activities. Through my own industry experience, I came to believe that a sustainable and constructible design is generated only when an engineer truly understands the situation in the field. Therefore, since class inception, forty hours of work on the construction site has been a requirement.[3] In addition to their in country construction work, students perform their design-project needs assessments, meet with their local project "clients" and benefactors, as well as with local design professionals and city officials, and gather design data. Therefore, time in country is divided between a construction site and gathering design-project data.

Students return to campus for the fall semester, and, in addition to their scheduled full fall-semester credit load, they complete the second three-credit course. They produce design-option feasibility studies culminating in an engineering report recommending a final design. We assume the international "client" accepts the students' recommendation. Students then develop construction contract documents as would be prepared by a design firm within industry. Students present their solutions on campus as if they were presenting to their international clients. Reports and contract documents are then sent to the international clients and interested parties. In the past, presentations have been filmed, translated, and given to the clients. Figure 8.1 shows coursework sequence and timing.

BUILDING A "BUSINESS" STRUCTURE

It is my belief that a capstone course should imitate and serve as a bridge to industry. As a result, I found myself coming from industry, pulling students away from calculations and theory and providing real-world experiences that better prepare them for industry. In industry, it will be rare if they use 5% of their traditional coursework in a typical day.

Drawing on my construction experience, I structured the ISD course to emulate the business of a design/build firm, a business environment students usually transition to within six to twelve months after the class experience (See Figure 8.2). This structure lends "positions" to multiple disciplines. For example, the marketing and communications department of the class is "staffed" by scientific and technical communication students.

Unlike in industry, at the pre-trip class meeting, students themselves form self-administered design teams of three to four, directed by team-elected project managers. As in industry, each team works on a different design project. Mimicking business activities, each student keeps time sheets and writes a daily learning log, which I view as a variation of daily contractor logs. Students note

---

[2]Fuchs and Mihelcic, "Engineering Education for International Sustainability," 2006; Mihelcic et al., "Integrating a Global Perspective into Engineering Education & Research," 2006.
[3]Hokanson et al., "Educating Engineers in the Sustainable Futures Model with a Global Perspective," 2007.

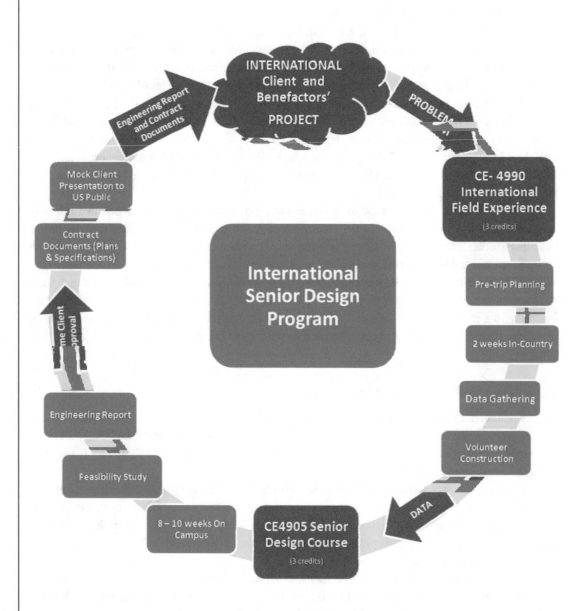

Figure 8.1: Example of coursework and time requirements for the International Senior Design (ISD) Program at Michigan Technological University. Both courses are summer credits.

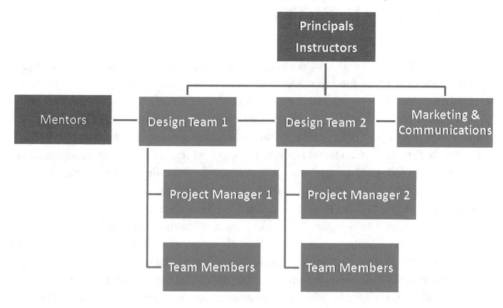

Figure 8.2: ISD Class Organizational Chart.

or react to gathering data, working on the construction site, learning local tools or techniques, team building, developing interpersonal relationships, and utilizing communications.[4] To further the industry correlation, I evaluate students with a 360-degree review of participation, attitude, and effort—components vital to team work.

In my opinion, the most important aspect of a capstone design course is that students experience the evolution of a project in an industry situation from project conception through completion of final design drawings and specifications. Learning this process is more important than the project type undertaken (structures, transportation, geotechnical, etc.). In addition, ISD students learn about a different country and culture, and, more importantly, themselves. US students generally arrive as strangers, learning to live with one another and work through culture shock and disappointments while developing project solutions in new settings. The shock of experiencing new sensations in an unfamiliar place impacts their drive, their performance, and their person. I am regularly heartened to see them develop a stronger and deeper attitude of wanting "to do it right," and a greater sense of citizenship and social consciousness than the senior

design students I taught on campus.

[4]Phillips et al., "Interdisciplinary International Senior Design," 2007.

FULFILLING THE NEED FOR A CAPSTONE COURSE, AND THEN SOME

Of the nearly 175,000 American students that studied abroad in 2004, only 3% were engineering students, which is only 1% of the US students enrolled in engineering programs.[5] In contrast, ISD has averaged nearly twenty-one students a year, or roughly 15-20% of civil and environmental seniors. By the end of 2008, 167 students constituting fifteen classes had accepted the ISD challenge, shaping the program into Michigan Tech's most popular and longest running international program.

I typically organized two classes per year since course inception. To assure adequate student enrollment to cover "summer salary," and at the request of the department head, in 2005 ISD students and I began marketing the class to freshmen civil and environmental engineering students so they could plan their schedules in advance. Three years later, we saw the results of this campaign at the 2007 informational session for 2008 ISD classes; with more than sixty interested students in attendance, the seats filled up quickly, and soon there was standing room only. The classroom was overflowing, with students listening in the hallway. The typical two-class sections thus grew to three course offerings in 2008 with the promise that additional instructional help would be found. Without any marketing for summer 2009, five or more sections would have been filled. The explosion of student interest we predicted had arrived.

Impressively, 51% of ISD civil and environmental engineering students have been female, which exceeds the 28% female enrollment in Michigan Tech's civil and environmental engineering department and is more than double the national average of 21%.[6] Astoundingly, 33% of the 2008 civil engineering female seniors at Michigan Tech enrolled in ISD. Course popularity among women is not the only positive outcome. At the completion of the 2006 courses, every female student stated that she now knew that she had chosen the right career and could see how to use her engineering skills to make a difference.[7]

THE QUEST FOR UNIVERSITY SUPPORT

Beginning with the first class, the verdict was in from students and industry representatives: ISD is a valuable, life-changing experience that effectively aligns with industry in many aspects. Despite the positive feedback ISD garnered, it received little recognition within the department. As the department head admitted, it was serving primarily to satisfy his need for additional capstone course offerings. With the increasing student demand for ISD, I found myself in search of broader support, including funding and instructional support, from the university. Sometimes Michigan Tech answered our calls for support, but other times we became self-sufficient and took ISD's needs into our own hands.

---

[5] Loftus, "Cream of the Crop," 2007.
[6] Phillips et al., "Interdisciplinary International Senior Design," 2007.
[7] Phillips et al., "Interdisciplinary International Senior Design," 2007.

MANAGING CLASS SIZE AND CREDITS

Although it would prove difficult to gain university support, thankfully the department supported some important aspects of the course. With increased student demand, it would be easy to increase the size of each class section, which we did. However, class size has an impact on my effectiveness in managing the in-country safety and class experience, so the department limited class size to twelve students. In my experience, three to four students per project team is optimal because some students tend not to fully participate in larger teams. Our goal is to provide each team a unique project, which both enhances student ownership and responsibility and increases our management responsibilities. In addition, the capacity of in-country accommodations and transportation constrain class size, as does the impact of uncertainties such as bus strikes, marches or other civil unrest, weather, and injuries or other health and safety issues.

Capstone senior design courses in civil and environmental engineering at Michigan Tech are typically three credits. The department head was concerned initially that ISD students would not meet expected classroom and study hours. It became apparent, however, based on ISD time sheets, that students were surpassing his expectations and spending much more time on their projects than the typical on-campus capstone student. In fall 2004, the department head increased the ISD course to six credits: three each for the in-country and on-campus components. What remained was a tremendous workload for students upon returning to campus. This second course was scheduled as summer credits to accommodate my "summer position," yet students needed to register for a full course load in the fall to meet full-time student status and financial aid requirements. Students struggled to find time for seemingly meaningless rote textbook problems while sinking heart and soul into projects that had become a part of them, projects for people they had grown to know and love.

STRUGGLING TO BREAK EVEN

As students' demand for ISD continued to increase in subsequent years, it was obvious that funds were needed for instructor salaries, equipment, computers, in-country assistance, and travel. ISD and my "summer" salary at Michigan Tech were completely funded by student tuition and lab fees, which include the cost for in-country construction materials. I strove to break even to minimize student expense. Much of ISD's in-country support to date was volunteered.

With the initial class proposal in 2000, and almost annually thereafter, I proposed approaches for corporate support. I began with a list of possible donors and approached various university development officials or grant researchers, each of whom would say, "This should be easy." They felt, as I did, that ISD's female involvement, international, cross-cultural, and interdisciplinary aspects, as well as humanitarian efforts, could be easily sold. As of today, nothing materialized other than intermittent funding from ISD alumni. Early in the program's development, university officials were hesitant to solicit small donations for such a specific program, fearing this could jeopardize larger departmental or university gifts. Soliciting funding from industry is difficult because, unlike more traditional connections such as mechanical engineering's tie to the auto industry, we do not have a singular industry client or sector to fund the program. ISD students are sought by many industry sectors. My industry experience, however, helps me see the added value ISD can provide students and how the program could be sold to industry or alumni partners and donors.

Perhaps my funding efforts were unsuccessful because my background was in industry and not academia. I do understand why such ventures are risky, and therefore unappealing, to university administration. Not only is it a nontraditional program, but it does not generate research funds. Moreover, there is tremendous risk in offering programs in an underdeveloped country, but everything comes with risk. Increased health and safety concerns are par for the course in a class such as ISD. Although precautions are taken and safety is instructed as paramount, students increase their risk of injury working on a construction site in a foreign country, where construction methods differ from those used in the US. They travel by foot through unfamiliar neighborhoods—the streets of which are frequently soaked with wastewater—and ride with unknown taxi drivers. Such risks are not encountered in the locality of Michigan Tech.

Another issue was my institutional position. With non-tenure-track status and a work schedule bringing me to campus only a few weeks in the spring and two months in the fall, I did not have the visibility of full-time tenure-track faculty. I was not surprised to find it difficult to be recognized and heard or given assistance, technical resources, and budget support by administrators who must answer to the permanent faculty.

MEETING THE INSTRUCTIONAL NEED OF STUDENT DEMAND

Adequate staffing to accommodate student interest is a common university problem. ISD is no different. However, the scope of my work *is* different than in a traditional course where course notes are taken from the shelf and reused. Student questions, project searches, international relationships, lodging, and air reservations begin anew each year. My responsibilities entail managing and administering the overall program, as well as multiple projects and student teams—similar to managing my projects and employees in industry.

The in-country coordination and direction became overwhelming for me rather quickly. In 2003, my mother stepped in to assist me in-country. Her fifteen years of work on international construction teams made her a natural. She could step in if I became ill or exhausted. I was spending more and more time with students in government offices and meetings, so she managed the construction site.

Also, as the program grew, I became increasingly involved with on-campus administrative, marketing and funding activities. In 2004, with the addition of course credits and associated summer salary, my husband, Dennis, became my assistant, as well as ISD's co-director. Dennis brought over twenty years of industry design and construction management experience to ISD, a natural continuation of our industry management duties: meeting with clients to develop projects and managing them via our student "employees." ISD had become much more than one three-quarter-time summer position. It became a family affair!

To ease our workload, department chairs recommended "relaxing standards." They did not understand the project-client-versus-design-product cornerstone of the ISD firm. To stay in business and retain clients, a firm cannot relax its standards, as lives, health and safety are at risk, along with the ability to secure future projects. We realized that without firsthand experience, i.e., without "going there," it is difficult for department managers to understand our position, our personal commitment and emotional investment in our students, the uniqueness of the program, and the slow pace in forming relationships with overseas community partners. And without interaction with the students throughout the courses, it is nearly impossible to comprehend the demands or the impacts. We began using the phrase, "If you don't go, you don't know." [8]

To accommodate the growing number of students electing ISD, additional instructors were necessary. Like staffing in industry, finding and hiring qualified instructors to meet student demand is no easy task. Our belief is that a qualified instructor should have sufficient industry experience, because industry background ultimately enhances the students' capstone experience. Yet it is difficult to hire such individuals given the research pressures of university departments. In the past, young faculty members without industry experience or tenure were those most interested in teaching ISD classes. They have been discouraged, however, and rightly so, due to the belief that the time commitment is too great to allow development of a research and tenure plan. We were unsuccessful in our attempts to add committed instructors to meet increased student demand while maintaining consistency in the program.

Consequently, the department preferred to limit student enrollment in ISD rather than aggressively investigate ways to expand a program for which demand rapidly increased. I see this as contrary to good business thinking. When a business has a product or service that appears to be a "good seller," it "invests" to meet the demand. With the new demand for global engineers and the standards set by universities requiring international experience, I believe it was time to invest.

"STAFFING" MENTORS IN AND OUT OF COUNTRY

After the realization that I alone could not physically and emotionally maintain the 24/7 responsibilities of living with the students in-country, arranging transportation, caring for the sick, coordinating with locals, counseling students, answering project-approach or design questions, and managing teams and projects all at once, help came not a minute too soon. When the department's student advisor and another former Michigan Tech alumna requested to accompany the second class, I quickly and gratefully accepted, not only because I knew I needed help, but also because I value others' perspectives and the wisdom of industry to infuse real-world experiences and values. While both were practicing engineers, civil and geological, respectively, the advisor's interest was in providing future student guidance, and the alumna conducted communication exercises and mentored.

With the third class began a flow of ISD alumni mentors that has not ebbed since. Student ownership of their design projects has been so great that 20% of ISD alumni have returned in-

---

[8] Phrase coined by the D80 Center at Michigan Tech (established 2007), the service-learning umbrella organization to which ISD belonged.

country to help as mentors and class assistants, paying their own way. Of those, 40% volunteered more than once. Although other industry mentors, who were not ISD alumni, also traveled with the class, ISD alumni mentors more easily understand student needs and reasons behind class rules or course requirements. In addition, they provide ISD tradition in a way that others and I cannot.

I require mentorship from industry practitioners for the on-campus course due to the number and variety of design projects. Practitioners volunteer their time and talents to mentor from afar, coaching students on project delivery and procedures, as well as technical concepts, before students go abroad. They then serve as functional and design experts to augment class activities, holding weekly meetings or conference calls to provide design feasibility advice and design review. The wisdom, experience and intergenerational exchange mentors provide are both helpful and often different than advice given by on-campus professors. When traditional approaches are not an option, it appears that students and industry mentors are more creative in thinking outside-the-box to develop alternate methods of obtaining data or approaching the design problem.

GOING INTERDISCIPLINARY AND INTERGENERATIONAL

With the third class offering, I was approached by students outside civil and environmental engineering. The first was a geological engineering student, followed by high school teachers obtaining their master's degrees and students in scientific and technical communication, business, and mechanical and biomedical engineering (See Figure 8.3 for the distribution of student majors.). I knew from industry the value of multiple perspectives, but I began to second-guess myself: "What will Angie (a geological engineer) really do?" I was reassured by experiences in the field.

I learned that the benefits of interdisciplinarity are realized regardless of students' home disciplines. I have been most encouraged, for example, to see the change in how engineering students approach data gathering and risk communication when scientific and technical communication students enroll.[9]

Although ABET requires multidisciplinary teams and open capstone course enrollment, other engineering departments and some civil and environmental engineering faculty have been resistant to the nontraditional teaming in ISD. Still, other departments desire to develop their own ISD. Including students from outside majors as part of multidisciplinary ISD teams allows more flexibility for project offerings and for professors to deliver the courses in a cost-effective manner—it is more economical for students from multiple disciplines to participate in the established ISD than for professors in each discipline to take classes composed of only one major abroad.

Limitations of our time in conducting the on-campus ISD classes led me to seek help from university professors outside my discipline. My first request was to a Spanish professor in 2005 to translate ISD reports for our Bolivian clients. The professor liked the business organization of ISD, so we agreed that ISD would hire the "*Empresa de Traducciones*:" the Spanish for Special Purposes course. Students began summarizing and translating ISD reports. One Spanish student later commented, "[T]his has probably been the most worthwhile class I have taken in my educational

---

[9] Phillips et al., "Interdisciplinary International Senior Design," 2007.

ISD MAJOR	Total Students	Percent
Civil Engineering	114	68%
Environmental Engineering	34	20%
Mechanical Engineering	3	2%
Geological Engineering	1	1%
Scientific & Technical Communication	3	2%
MS in Education	5	3%
MS in Engineering	3	2%
Biomedical Engineering	3	2%
Business	1	1%
TOTALS	**167**	**100%**

Figure 8.3: Distribution of Student Majors (2000-2008).

career. The most worthwhile aspect about this project is that we are making an impact on the lives of people halfway around the world."[10]

To manage the ISD firm's communications, I approached a professor of scientific and technical communication (STC). This yielded structured STC class assignments for promotional and marketing materials, grant writing, and audience analysis. One STC student later commented, "I learned so much about working with other disciplines for the success of a project. I could see what we writers and designers could offer engineering projects, and that was awesome."

As a result of these interdisciplinary classes, interested communications and Spanish students enrolled in ISD and continued their disciplinary studies with projects that benefit ISD. 20% of the first "*Empressa de Traducciones,*" regardless of their major, took or signed-up to take ISD as a result of this Spanish course. The humanities department approved ISD for their language minor and certificate programs. In addition, risk communication class assignments were based on ISD projects, and a documentary film class has used footage shot by a Bolivian university communication student for their class project. For me, these humanities classes contribute to my dream of interdisciplinarity

[10]Boschetto-Sandoval et al., "Pilot Collaboration and Program Development," 2007.

and international exchange. I also find it ironic that support and respect from these professors outside of engineering have been greater than what I received from my own department.

I have also found that participants in ISD projects can come from surprising sources. When teaching a master's level "engineering for high school teachers" course at Michigan Tech, I was asked to give an after-class presentation on ISD. In this master's program, teachers learn how to incorporate real-life engineering into math and science classes and complete a six-week engineering internship. Four teachers approached me to request using ISD as their internship. Their program director was open to the idea, but the two-week duration posed a problem. To help them fulfill their responsibilities, I arranged for the teachers to teach physics and English in local Bolivian schools for two weeks prior to the ISD course. Also, by being part of student design teams in ISD, the teachers saw firsthand the development of an engineering project from an idea to construction. Some now use similar activities in their US classrooms. And one teacher has actually taken his own high school students to volunteer and interact with students at the school where he taught in the Bolivian jungle. Other precollege teachers are now working with me on a precollege ISD program.

BUILDING IN-COUNTRY RELATIONSHIPS

When it rains, all streets are flooded, kids can't attend to school. People take off their shoes to cross the streets. Lots of people get fungal infections and the typical diarrhea and vomit in children. [The ISD students working on projects] is a wonderful idea. Based on their studies we can present a project and look for funding, so someday we can have sewers or paved streets. I appreciate the sacrifice and will of these students.
— President of Telchi Neighborhood, Santa Cruz, Bolivia

Although ISD started at a single school site in Santa Cruz the project quickly drew the attention of the *barrio* (neighborhood) residents. Soon local projects spanned several neighborhoods, and we began working with various districts of the city. The more projects ISD completed, the more project requests arrived from various *barrios*. Personal relationships and the building of trust were key.

Beginning with the second class, school officials in Santa Cruz saw new project possibilities and expanded their requests to include building design analysis, wastewater systems, and other methods of resolving flooding. The latter led us into the *barrio* to define the watershed for ourselves and, in response to a spur-of-the-moment request from a few neighbors, to examine their flooding. The experience garnered mixed reviews [from barrio officials] because we were not prepared to springboard into the neighborhood. Also, neighbors were not fully aware of why we were there and what our goals were. A small contingent of neighbors did understand and took us to the city maintenance department, which resulted in a donation of earthmoving equipment. Many, however, were suspicious of us.

The project carried over to the following year, when a new team worked intimately with the barrio president and members, who had no idea how to solve the flooding problem, to better identify and understand their concerns. This approach was so effective that neighbors took the students to local governmental offices, which resulted in the gain of invaluable project resources, including loaned equipment and a donation of reinforced concrete pipe, and support from the sub-mayor. The neighbors' initiative was so great that the survey, engineering, and construction of the project were nearly completed before the students returned to the US. Building support and personal relationships with local neighbors opened doors into other government offices, and since then, we have had their support in identifying and prioritizing ISD design projects.

Although some officials worried at first that ISD would take work away from local designers or city staff, it became clear within two years that ISD students were providing design support for projects that would otherwise be left undone. Students also come to understand they are in-country to address problems that would otherwise go unmet. Respecting the value of local knowledge and the use of local materials, students now work closely with city engineers and officials, who also coach local design methods. Word has spread, and project requests now come from central governmental engineering offices and other government districts in surrounding areas. In 2008, ISD and the central city government collaborated on a canal project that was too large for one ISD team alone, so the city provided surveyors to work *within* the ISD team. Although it took three years to build this trusting relationship, ISD students and Bolivian surveyors continue to collaborate on projects.

During the first four years in Santa Cruz, I had in-country language support, as well as support for group transportation and logistics. Due to death and retirement, however, I was left on my own. By that time, I was well connected with the city district —or so I thought. Local elections quickly changed that, as my collaborators disappeared! I began anew, visiting the new officials, presenting our past projects, re-cultivating relationships, friendships and trust, building upon my former network. During one class, a resident from a neighborhood in which we were working suddenly told us we needed to attend a school ribbon-cutting ceremony. We were whisked away in a "hijacked" city bus to the unfamiliar school. We waited for quite a while, and students complained we were wasting time. Suddenly the mayor approached us, introduced himself, and offered to provide valuable project resources. The local resident knew all along this chance meeting could be fruitful. I learned to trust relationships, for unexpected methods sometimes yield unexpected results. I have also learned that regardless of my preparation and organization, virtually everything in-country seems to be impossible until the last minute! Our in-country relationships are invaluable to our success, but we must be persistent and patient at the same time. I push my limits as I expand my circle of contacts, looking for projects in fields new to ISD, such as biomedical engineering and STC.

Furthermore, there are project trade-offs to be balanced, as illustrated in a water-distribution-systems project in the Dominican Republic. In writing their feasibility study and engineering report, ISD students wrote about the advantages in the locals' request for water taps closer to homes so villagers and children did not have to walk a half-hour to fetch water. Students surmised this allowed extra time for children to attend school. I reminded them that the small one-room school only held so many students, and, as is customary in many countries, they only attend school half-days. What would the children do with the extra time? "Study," my students replied. When I challenged them whether they would spend more time studying if they were in the children's shoes, the answer was, "No!" So how could we expect local children to choose study over play?

Following my students' response, I shared the story of Chickatica, who was in her eighties and was happy to learn about the possibility of having water piped closer to her home, because then she would not have to "depend on the *muchachos* to bring her water." I felt a stab to my heart when I first heard this. The students in the project were so thrilled to be "helping" these people that they could not see the social and intergenerational impact of the project. With water closer to their homes, per the villagers' own requests, we were helping remove a social responsibility of the *muchachos* and a means of intergenerational communication. With every change for the good comes the not-so-good.

The lessons of the not-so-good have a lasting impact. An alumna recently reminded me of this, as she relies on Chickatica's story in every interview she conducts. Another, at his own wedding, said this particular lesson affects his daily engineering decisions. The stark difference between their worlds and Chickatica's still affects them after seven years.

From this experience, we have learned that in order to completely weigh the pros and cons of a project, we must carefully consider the cons that are not readily apparent—what is considered to be development by us, or even those in a host country, can be a double-edged sword, as is demonstrated by Chickatica's story. As I have said previously, everything comes with a cost—is it worth it?

Finally, ISD has generated interest among engineering students in its host countries, especially Bolivia. Beginning in 2002, civil engineering students (with English language skills) from two universities in Bolivia asked to be included in ISD. I granted their request with some trepidation. What I found was that they too wanted to gain field experience not offered at their universities. They skipped their own classes to help with projects. Participating in the ISD project exposed them, for the first time, to underprivileged living conditions in their own country.

Forging these new relationships and building trust takes time; however, a leap was made in 2008 when a Bolivian department chair participated in an ISD class. The visit yielded a possible collaborative ISD class at UNIVALLE, melding Bolivian and US students under the direction of Bolivian and US instructors. Although there is much excitement, it will be a lot of work to physically move the program to a new site.

RECEIVING INDUSTRY AFFIRMATION

Your programs are great! I no longer look at a resume unless it has a significant internship or an experience such as [ISD].
                    —Comment from the director of a leading engineering consulting firm

When the first ISD class returned to campus from Bolivia and presented their design solutions to an audience of professors, interested students, and the departmental advisory board, two of the students actually presented the class project. However, all twelve stood to answer questions, each stepping forward and responding admirably to the audience's questions in an unplanned fashion. When water resources professors asked detailed questions about hydrologic data and calculations, the students explained their unconventional approach. Without a sound understanding of the locality, the professors were not open to the outside-the-box analysis, and leaving the presentation room, tearful ISD students were criticized by professors and student peers for having done an imitation of Habitat for Humanity. By contrast, industry advisory board members commended the students for their approach and methods, and commented that they preferred to hire students with experience who could rely on basic theory and unconventional approaches, rather than traditional students with higher grades.

As we have learned in our experience, the industry acceptance and support garnered by ISD has contrasted greatly with the typical feedback given by faculty. Too often university faculty will write off students for thinking outside the realm of conventional theory, dismissing them for not having taken a course and, in their view, lacking knowledge in a particular area of engineering. In contrast, industry recognizes the worth of independent critical thinking and design. Practicing engineers look at problems from multiple perspectives to come to a solution, even if it is nontraditional. Unconventional classrooms, such as an underdeveloped country, necessitate unconventional thinking. ISD students are required to take into account many factors, such as local practices and culture, to design a functional, sustainable solution, oftentimes employing methods that are novel to the US.

Industry recruiters consistently seek ISD graduates over traditional senior design students because they have handled demanding work, and they can think outside-the-box and understand an engineering project from inception through construction. Recruiters have pressed me to meet "the rest of the ISD students" because they had talked to a few, realizing they were passionate and hardworking—some search out Dennis and me at a career fairs. Students have been hired on the spot, being told that this is not typical corporate practice, and others have been told that their ISD experience was more valuable than a co-op experience, as they were required to do more in ISD than they would be allowed to do on an internship assignment in the US. Surprisingly, more than six years after graduation, an alumna had her resume professionally reviewed. As she confided, "He ripped it to shreds and took much of my industry experience off, but told me to never ever remove my ISD experience."

WHAT IS ISD WORTH?

CAN LEARNING OCCUR IN TWO WEEKS?

There is no doubt that one cannot truly learn about a culture in two weeks. Although my students must interact daily with the locals across ever-present language and cultural barriers, none of us can fully become assimilated into the culture, even after years. Since our skin color, accent, and cultural origin will always be apparent, locals will look at us differently. We will always be treated as guests because that is the cultural norm. For example, on the Dominican Republic job site, ISD students worked with the local construction workers to haul and place block and concrete. They viewed us as the guests, so whenever we grabbed a trowel, they took it out of our hands.

Still, nothing is equivalent to spending time abroad. No amount of research beforehand can equal the understanding of culture and processes gained firsthand; you must *go* to *know*. An alumna reflected on the importance of this during project design: "You couldn't just move the earth and dig ditches like you would in the US because you might disturb the city soccer field [neighborhood crossroads], and if you did that, it would be worse than flooding their homes every year. You definitely have to learn about values, cultural concerns and working conditions before you hope to accomplish a project."

In two weeks, students cannot fully appreciate the project generation process or the political or municipal funding issues; however, they quickly come to an understanding that these, as well as other social and cultural issues including health, community participation, and gender, play a large part in their projects' success. They realize that the communities could do these projects themselves if they had the resources. Understanding these issues is probably more important than the technical background one has, and such appreciation *can* be gained even in the two weeks of ISD. Regardless of the amount of study or time spent in-country, neither I, nor the students, will ever fully understand the culture. While this may hinder our work, it simultaneously provides opportunity for cultural growth and understanding for us and our local counterparts.

STRETCHING COMFORT ZONES: THE EFFECTS OF DISCOMFORT

Before I left the US, I talked to past students, looked up photos online, and stretched my imagination. Even though this preparation may have lessened the shock, I was still left speechless. The bathrooms were a swamp! The toilets didn't flush and all had waste piling up in them. The seats were caked with dirt and flies while the tanks seeped liquid onto the floor. Outside, sewage was coming up from the ground. Wet spots surrounded the septic tank and muddy puddles were adjacent to the bathrooms. The children had footpaths that ran straight through these infected areas.

—ISD Student Learning Log

Research shows that the further a person is pulled from their traditional comfort zones—their culture, language, daily routine, food, pastimes—the more impactful the experience.[11] During the abroad portion of ISD, US student comforts are traded for tight schedules, early curfews, limited cell phone and internet use, roosters crowing all night, doing laundry by hand, no TV, and the milk lady squawking though a bullhorn before dawn.

In my opinion, the secret behind ISD's impact and success is that students are taken far outside their comfort zones. How can the above reality not jar the "comfortable" US college student? They may be designing a storm or wastewater solution similar to one in the US, but they are situated in a culture foreign to them. Rarely can students communicate fluently in their host country's tongue, and they have the opportunity to interact and build relationships with locals and project benefactors who represent a range of economic, social, and political positions that are underrepresented within their comfort zone.

The potential social, cultural, and political impacts of projects in the ISD setting are glaring compared to those in the US. Dengue, malaria, garbage, storm water mixed with untreated

[11] Cumes, *Inner Passages Outer Journeys*, 1998.

wastewater—these aspects are now impossible for students to ignore. How often do traditional cap-stone students move beyond meeting with the US city officials to deal with the impacted neighbors and benefactors to appreciate their stance? One student learning log described the importance of considering social factors in engineering,

This experience highlighted the need for modern engineers to be aware and concerned about much more than simply the engineering aspect of a challenge. They must take into consideration social, political, financial and cultural factors when creating a solution; the ability to integrate and engage local people into their own engineering project is critical not just to the engineer's success but more importantly for the project's long-term success. Engineers cannot work in an environment isolated from the people a project impacts. Social consciousness in engineering is a must.—ISD Student Learning Log

ISD stretches comfort zones not only in language and culture, but also in construction work with local residents. Many students have not done much physical labor before ISD, and job site tasks test them. They quickly appreciate and respect the construction workers' strength and stamina as well as their wisdom. Students quickly learn that ISD is not a vacation. The discomfort and disruption of their routine and the removal of US amenities add to the impact of the experience. Like a wilderness experience, ISD is short, intense, and uncomfortable; one where you find out who you really are and what skill set you possess. One of the students' biggest gains from experiencing and adapting to the discomfort and their project work is tremendous growth in personal confidence. From my perspective, students provide critical help in addressing problems that would otherwise go unmet, but the experience is so shocking, so enriching that US students will be affected more than the host country. A school's walls may be constructed, a wastewater system designed, but the grander value is to the US student. Students learn the value of family and friends and realize the insignificance of material items. They see a more genuine patriotism and a true respect for elders and titles. If they pause for a minute and remember they are members of a much larger global community—then that is a value of the program.

The ISD experience opens students' eyes, hearts, and minds while allowing them to under-stand their discipline and civic responsibility, both personally and professionally, as it relates to real social needs. It marries social consciousness with engineering, more than traditional coursework. In fact, ISD alumni, commenting on their outlook toward corporate clients, clearly see themselves as more perceptive and receptive to client needs and social, cultural, and political issues than their co-workers.[12] And since the projects truly become part of the students, they want to see the projects come to fruition regardless of timing.[13] An alumna/mentor commented on how her project still possesses her,

I sit here, 3 days before attending [ISD student presentations] to watch my ISD students with pride [as a mentor], considering the possibility that there is still hope for my own ISD project to come back to life. The impact of ISD hits me once again. After 4 years,

[12] Wright et al., "International Senior Design," 2008.
[13] Phillips et al., "Interdisciplinary International Senior Design," 2007.

life changes, and a blooming career, my [ISD] project is still with me and always will be. Where else does a senior design last a lifetime?—Former ISD Student email

I feel that students find an identity within ISD—a new "comfort zone." A student wrote, "Just as many past ISD participants have told me, 'If you don't go, you don't know,' rings truer now that I have gone. It no longer feels like a knock on those who have not gone, but an affirmative message for those who have gone."

This is why we continue; why we persist: for the impact on the US student. Nearly every student finally "gets it." It is true that almost every ISD student has commented verbally or in learning logs about the quantity of work, strict rules, tight schedules, sore muscles, sickness, or other frustrations; however, only four or five of 167 have stated that they wished they had *not* done ISD.

One explanation for this response may relate to students' expectations. One student said, "I expected [ISD] would be a little less work, a little more fun (sightseeing)…I think that needs to change a little bit, especially if we're spending all this money. It's exhausting." The student felt that since he was paying for the experience, the class should conform to his wants and expectations—that since he was a consumer of the class, he should be satisfied with the "product" (experience). He reacted adversely to being taken out of his comfort zone—partly because he could not speak Spanish fluently. He said, "Spanish made it difficult. When I was with the translator, I was fine, happy, *comfortable*. [emphasis added]" Although he did not believe ISD was "worth it" for him, he acknowledged that, "for some people it is, and that's fine for them."

However, even those students, including the aforementioned, who said they wished they had not participated also said in the end that they indeed gained valuable lessons, and they too were thankful, as evidenced by the following student comment, "No matter what negative feelings I have about my project or the program, they pale in comparison to the amazing educational, emotional, and professional experience ISD is. It is easy to talk about the negative nitpicky things but the unexplainable overall character growth is what truly matters." The long days and sleepless nights are lost, and the good memories and impactful results remain—and students tell me they will remain for a lifetime.

IN CLOSING

Looking back, it is easy to see how this grassroots development fits broader interests in engineering education reform: increasing diversity, attracting new students, and training a new generation in sustainable practices. Students pushed the status quo and demanded education with a different approach, one that could make a difference not only for those in the developing world, but also for themselves.

I developed ISD using the resources I had available, building on our industry experience and beliefs. It is an unconventional program with unconventional faculty, with strong work ethics, values, and expectations. Yes, I had the model of my volunteer trips, but ISD is more a by-product of immediate circumstances and our systematic hard work. I did not knowingly design the "comfort stretch" into the course; that just happened. After the first class, I realized that it is the surprises, the welcome and the selfless acts of the local people that stir the US students to action. I cannot orchestrate these. How could I predict that on our last night, soon after the loss of his father, a young boy with a small brown paper bag would change our lives by giving us each a "gift from his bedroom" that we were to "keep close to us and never give away." Some of those gifts, he told us, were treasures his father had given him. Nor could I plan for the soccer player who, on leaving, takes his precious soccer medal from under his shirt and hangs it around a shy ISD student's neck because they "connected" though soccer. Later, I find the shy student alone, sobbing uncontrollably and with coaxing he tells me that the local people have been kinder to him than *anyone* in his whole life. Incredibly, both of these events took place the *same* evening. These are the moments, the gifts, the awakening, that are associated with the "class," and are beyond my control. These are the moments that make ISD….they are the magic!

Via the ISD program and its students, we intentionally infused social responsibility through international engineering education into the heart of the university. We came from industry to bridge students from theory to the "real world." Instead of bringing our industry experience *into* the classroom, we now bring the classroom *into* the real world of industry—to "professional practice" in a developing world setting.

Students, especially females, initially enroll in our programs "to help others" or "to make a real difference."[14] To see an explosion of female engineering students that overwhelm our demographics excites and encourages me. It is clear that solving engineering problems of society and environment at the humanitarian and ecosystem level attracts women to engineering. A 2006 female student said,

> I participate and do very well in my engineering classes. But I am never satisfied with these classes because. . . I do not feel I am contributing to the world in any positive way. . . . I understand that I may have needed many of my previous classes in order to effectively participate and that the mental discipline I acquired through them is important. But the truth is I value the experience gained in ISD more than I value every engineering class I have ever taken combined, because it involved something that is real and brought face-to-face at every level with all of the humans involved.— ISD Student Learning Log.

It pleases me now, after years of hearing the new male students introduce themselves with the desire to "travel" or "explore," to hear the current male students saying that they now want to "improve their employability," "give back," or "make a difference." I am grateful for this change in attitude and hope male participation within ISD will grow.

I firmly believe ISD courses are a model for breaking the gender gap, and the long-established and well-built departmental boundaries of our universities. If students are to appreciate interdisciplinarity and diversity, we, as educators, must as well; and we must *lead* the way. *This* is why I continue to work with ISD.

How has this changed me? I see things differently. I taught with passion in the classroom but this is *more* passionate; *more* intense. I am firm in my beliefs about international and sustainable

---

[14] Phillips et al., "Interdisciplinary International Senior Design," 2007.

education and the importance of hands-on experience. It is critical to the future of our planet and our peoples that we include sustainability and constructability into *every* course, marrying it with the social consciousness that is mandatory in "good" engineering. And finally, I *know* and have *lived* the difference substantial industry experience can make in the university classroom when it is delivered with care and passion.

Like my students, I too have been thrust outside my comfort zone and challenged to critically look at how and what I taught, to learn a language, and to see the relationship of societal and environmental needs to engineering. Similarly, to my students, my eyes, mind, and heart have been opened. Reflecting back, even with time commitments, the travel demands, maintaining international relationships, and the emotional investment with students, I *cannot* return to industry, and I would hesitate to return to the traditional classroom. None of us can turn back now.

POSTSCRIPT

Writing the chapter gave me a chance to reflect on the development on my trajectory, or personal geography. In 2008, during the review of this chapter in Washington DC, after questioning by a reviewer, I realized I may never receive the university assistance I asked for because in the past I persevered without it. In other words, had I trained the university not to listen to our pleas? I was exhausted. The program had outgrown us and our capabilities. There were five full ISD classes of students who had class down payments in-hand waiting for us. The flight home from the workshop was filled with further self-questioning. Dennis met me at the airport and said, "We need to talk." We'd both come to the same conclusion at the same point in time. In the past, we had each been there at different times. Now we were there together—and independently. The next day, we resigned. We had nothing—no other employment. However, other schools had approached us in the past. We had faith that interest remained, but the question was: would they see the value of a non-traditional program, offered by non-traditional faculty, in a non-traditional manner?

ACKNOWLEDGEMENTS

This chapter is written with gratitude to those who helped shape the ISD program, but especially the ISD students and our global partners, as without them, there is no ISD. I also lovingly thank my husband, my International Senior Design Co-director, Dennis Magolan, for his assistance and vision and my mother, Marilyn Phillips, for her unwavering assistance and support. Thank you also to Karina Jousma and Lucas Shanks for insight on the text.

REFERENCES

Boschetto-Sandoval, S., C. Sandoval, and L. Phillips. "Pilot Collaboration and Program Development: Engineering Senior Design and Spanish for Cross-Disciplinary Literacy." Paper presented at the 114th American Society for Engineering Education Annual Conference & Exposition, June 23–28, Honolulu, HI, United States, 2007. 187

Cone, Inc. and AMP Insights. "Cone Millennial Cause Study." Cone Inc. and AMP Insights, 2006. 175

Cumes, D. *Inner Passages Outer Journeys*. St. Paul MN: Llewellyn Publications, 1998. 192

Fuchs, V.J. and J.R. Mihelcic. "Engineering Education for International Sustainability: Curriculum Design under the Sustainable Futures Model." Paper presented at the 5th Annual ASEE Global Colloquium on Engineering Education, Rio de Janeiro, Brazil, October 9–12, 2006. 179

Hokanson, D.R., J.R. Mihelcic, and L.D. Phillips. "Educating Engineers in the Sustainable Futures Model with a Global Perspective: Education, Research & Diversity Initiatives." *International Journal of Engineering Education* 23, no. 2 (2007): 254–265. 179

Loftus, M. "Cream of the Crop." *ASEE Prism* (Summer 2007): 28–33. 182

Mihelcic, J.R., L.D. Phillips, and D.W. Watkins. "Integrating a Global Perspective into Engineering Education & Research: Engineering International Sustainable Development." *Environmental Engineering Science* 23, no. 3 (2006): 426–438. DOI: 10.1089/ees.2006.23.426 179

Phillips, L. D., M. Brady, and K. Jousm. "Interdisciplinary International Senior Design: How Service Learning Projects in Developing Countries Support ABET Accreditation." Paper presented at the 114th American Society for Engineering Education Annual Conference & Exposition, June 23–28, Honolulu, HI, United States, 2007. 181, 182, 186, 193, 195

Wright, H.E., L. D. Phillips, and J. R. Mihelcic. "International Senior Design: Assessing the Impact on Engineering Students after Graduation." Paper presented at the 115th American Society for Engineering Education Annual Conference & Exposition, June 22–27, Pittsburgh, PA, United States, 2008. 193

CHAPTER 9

A Lifetime of Touches of an Elusive "Virtual Elephant": Global Engineering Education

Lester A. Gerhardt

My life, like most I suspect, has been influenced and guided by thoughtful, planned individual choices I have made, coupled with my instinctive reactions to unanticipated circumstances and windows of opportunity I have encountered along the way. These generally uncontrolled circumstances, which may spontaneously be presented, may be either passed over or seized with a special passion. Perhaps it is deep insight, experience, education, or just plain dumb luck that a person effectively and successfully joins elements of these two groups together from time to time in one's life.

To use a metaphor, there is the classic story of several blind people touching different parts of an elephant, but without the ability to integrate the different parts and visualize the whole, they never realized that it was but a single elephant they all shared. As I was reflecting on my life and times for the purposes of this book, it was only then I realized that through a series of planned and circumstantial events, I had *touched* different aspects of globalization, engineering, and education throughout my life. But only through this personal geography did I more fully understand these were developmental parts of the same interest—my intellectual 'virtual elephant'. These *touches* helped develop, integrate, and support what eventually became my passion for globalizing engineering education as I pursue it today.

In this personal geography, I cite some of these encounters and tie them together to present a type of roadmap of my life. This path taken has caused my personal life and professional career to become inextricably intertwined to the benefit of both, and traveling it has taught me much about others, myself, and my engineering self. This introspective reflection has helped me better understand how the confluence of certain conditions and events brought me to the point of being passionate if not obsessive about the need for, and commitment to, making a global dimension part of our educational system and methodology. I hope it likewise serves to not only provide some helpful hints along the way for others embarking on a similar path, but also that it gives you pause for self reflection, motivation, and guidance regardless of the road you are taking in your life and career.

DIVERSITY ON THE STREETS

On the streets in the Bronx, New York, where I was born and raised, we learned mostly by experience, by doing. Scrapes and bruises were worn as a badge of honor, and accelerated the learning curve. We learned by doing and watching others, and simply picked ourselves up and tried again until we succeeded. We learned quickly and early in life, to know thyself, and also to read people and know their sensitivities, strengths, and weaknesses. We learned respect, consideration, caring, acceptance, and a number of other fundamental social and people skills, which proved to be critical to success in my personal and professional future. "Working well with others," as it then said on our Public School Report Card, proved not to be a cliché but a valuable asset. It is certain that my interpersonal skills are a direct result of such an upbringing and have boded well for me in my research, teaching, and administrative adventures. Today we call it team-building, communications, leadership, and entrepreneurship. We teach these in a more formal classroom setting in a type of virtual environment, but it is not the same as the real world environment on the streets where I grew up—a place and time where need proved to be the greatest motivating force, and education was understood to offer the solution to that need.

This *touch* with learning by experience at an early age (just do it), provided one basis for later emphasizing the need for and importance of a semester-long immersive experience abroad for our students rather than other simpler alternatives.

We were not taught diversity; rather we lived it. My neighborhood was truly mixed ethnically and racially, although dominated by those of European origin. Everyone's parent or grandparents had come from another country. An international perspective was part of our daily life, as we listened to their different languages, and they recounted tales of the "old country" for us. Everyone knew of each other's religious holidays, and in fact sometimes made them their own. It was much later that several of my non-Jewish friends acknowledged that they knew not going to school on Jewish holidays was meant for people who were Jewish, not everyone (I bet they knew all along). For many years, I visited St. Patrick's Cathedral on Christmas Eve with them and was truly inspired as we prayed as one people despite the fact no one understood Latin.

For street games, people were chosen because they were the best at that particular sport or activity, not because of gender, race, or religion. Most people around me were different from one another in many respects, but the differences of gender, race, ethnicity, or religion were not regarded as defining characteristics. Rather, differentiating characteristics were simply height (Tall Eddy), weight (Chubby Charlie, or Skinny Ronnie), or wearing glasses (Four eyes Francine). Quality was the main discriminator. Running the fastest, making the most baskets, or hitting two sewers in stickball was what got you chosen for the team. This first *touch* with diversity taught and instilled in me the inherent value that people from different cultures brought to our gang then and to the United States today. I did not need be taught to accept it since there was nothing to accept. These were simply my friends to be trusted, respected, and honored for who they were rather than what they were.

Perhaps this time was also my first touch with the concept of globalization, although we certainly did not know that word either. We were just friends that played, laughed, and learned together, without regard to our family's country of origin. Bringing this forward to establish relevance in today's environment, our government recently sought to classify some international students studying in U.S. universities as a "deemed export" based on their country of origin and thereby restrict their access on university campuses. Perhaps it would be beneficial for each of us to recall our own ancestry, upon which the United States has built its strength and attained its unique leadership position in education, finance, sports, media, etc., in little more than two centuries. How different is it to require selected students to display identification indicative of their background than it was for my ancestors to wear the yellow Star of David to identify those to be escorted to Dachau? Separate but equal was not right then, and it is not right now.

I also learned about special circumstances that could sometimes override reasoned judgment. In those days, sometimes if your mother baked especially good cakes, that might have swayed someone to choose you for their team even if you were not the best. Such special factors continued to come into play in later life, and it was good to learn about them early. I still always think about how a person in a leadership position got there. Was it sheer quality of performance, or did they marry the boss's offspring? Or maybe they are a good baker of "cakes" of one sort or another.

We learned to respect parents, not just one's own. We learned to respect and honor teachers and authority in general. Only in later years did I learn to fully appreciate all they did. Perhaps this first *touch* with education is the underlying reason I chose to spend most of my professional career in academe, seeking to give to others as my teachers had given of themselves to me. We respected each other as individuals, as well as each other's property, little as it might have been. There was no theft between those who lived in such close quarters. Apartment windows were often left open to catch a summer breeze. Despite being interconnected by a series of fire escapes, they were never entered by anyone, and the windows were never barred as they are today. Today, I usually complete a presentation on how to succeed in establishing global partnerships with insisting on TRC-Trust, Respect, and Confidence. I well remember when that first became important to me on those streets in the Bronx.

All these lessons have served me well, and I have always valued and sought to have my children and my students gain broad exposure to the diversity of their environments. Hence, exposure to diversity and education became my mantra at an age when I did not even know what these words meant.

VALUING AND PURSUING A COLLEGE EDUCATION

Most of all, on the streets of the Bronx, we learned to appreciate the value of education. Education represented the "light at the end of the tunnel," much as it does today, especially for other first-generation Americans. And today, if ever I give a problem on an exam for which I did not completely work out the answer in advance, I have only to select a first-generation American student to see how to do it correctly.

I went to the City College of New York (CCNY), one of the finest colleges. I was to later learn this was the school that claimed the largest number of undergraduates who went on for their Ph.D. and the largest number of graduates that became Nobel Prize winners. However, these important facts had little bearing on my going there or studying engineering. Simply put, it was very high quality and quite affordable. The tuition was free. I was young, and it was nearby, reachable by a short subway ride from my parents' apartment where I happily lived until graduation.

Those were not the days of filling out ten to fifteen college applications, to be followed up with campus visits to at least the top five to meet the faculty and the President, and check out the dorms. I was good in science and math according to tests taken in junior and senior years of high school, many times with perfect scores. In any case, my parents, like others of that time, felt if you pursued what you were good at, you would do well, progress, and be justly rewarded. With capability was presumed interest. So since I could do "it," I went down an engineering path. Actually, at that time, I never knew someone who was an engineer or really what an engineer did. But I had built a Knight Kit once, and that might have been the determining factor for me to study the most difficult discipline of the time, electrical engineering. Sometimes you do not need all the answers ahead of time. The expression, "excessive analysis can lead to paralysis" comes to mind. I just knew that I needed to get a college degree to get ahead, and that I could intellectually handle the difficulty of the work. I had yet to develop passion for my "chosen" profession. That was to come later.

I not only did well in math and science, but I also did very well in English. After I won a citywide English essay contest, one English teacher had written, paraphrasing Winston Churchill, "Never has so much been said so well about so little." I later learned this was not intended entirely as a compliment. The importance of good communications skills was a major factor in the success of my own professional and personal development. I also still write for pleasure and have been published in non-technical fields, although these have never found their way on to my resume. This early touch is the reason I promote well-rounded students and always convey the importance and fundamentals of verbal and written communication skills to them.

ADD A TOUCH OF BALANCE

For most of my summer employment, I worked as a lifeguard during the day and dance instructor in the evenings in the Catskills north of NYC. I met my first and only wife at that same hotel. We still watch the movie *Dirty Dancing* with a special set of memories and fondness almost fifty years later. And, yes, we still do dance whenever the music is played.

Certainly the major touches of my "virtual elephant" over these most formative years of my life were education, engineering, and an international awareness. But to properly integrate these together, a touch of balance was needed. My wife added that special ingredient—the appreciation of balance in life. I will never underestimate the lifelong support I have gotten from her, implicitly and explicitly. Without her emotional support, and without her handling the affairs of state in the family, I could never have availed myself of the opportunities that were offered to me, especially to travel internationally. She continues to remind me of the value of that balance even today.

My belief in the need for balance in life has deeply shaped my work in engineering education. The infrequency with which today's engineering students attend seminars outside their own specialized discipline is matched only by the infrequency with which these same students voluntarily travel outside the United States to expand their horizons and explore other cultures as part of their educational program. The importance of balance of life's components that my wife brought to me is the reason I seek a balance of multidisciplinary education and multicultural exposure for my students today. It is probably the root of my motivation for integrating education and engineering with an international experience.

A FIRST TOUCH OF ENGINEERING AS A CAREER

After graduation from CCNY, I took a position with Bell Aerospace Corporation. The choice was made as much based on its location and the nature of the offer, as it was on my knowledge of the company itself. That said, the decade I spent working there was inspiring, challenging, and very rewarding. I worked on the visual simulation of space flight including the earth orbital rendezvous maneuver, the trans-lunar trajectory, and focused on the Apollo Program's moon landing. These years helped me not only gain technical expertise, but it also broadened my knowledge and appreciation of the importance of multidisciplinary engineering, and the relevance of my chosen field of electrical engineering in this context. One could hardly be restricted to a single discipline while working on the space program. This industrial experience served me well in years to come when advising students. In helping them to formulate their career plans, I stress the value of experience before committing fully to an academic career.

Moreover, these years at Bell further solidified in my mind the importance of gaining an international perspective for the field of engineering. I worked with international scientists who were brought to the U.S. after WW II under Operation Paperclip. It reminded me of those days in the Bronx when you just picked the best player for the game at hand, regardless of their background. The objective of this game was much bigger of course: which team could get to the moon first. But the strategy was much the same. Each team would do almost whatever it took to win the race using the highest quality talent.

It was then I experienced a real awareness that the science and engineering workforce needed to be regarded as more than a national resource. I now emphatically feel that global issues can only be solved by using global resources, and that the interdependency of countries with respect to environment, energy, finance, and health of our planet and ourselves mandate cooperation rather than conflict. On the fine line between cooperation and competition, let us choose to err on the side of cooperation with peaceful long-term objectives.

I also came to understand the long-term futility of war despite any claimed short-term objectives. Members of the U.S. team were the very same people who designed the missiles and bombs that killed so many Americans and others only years before as members of another team. For many years, I worked directly for Dr. Walter Dornberger, Vice President of Research at Bell Aerospace. He had been responsible for the V-1 Buzz Bomb and the V-2 rocket development programs as well as directing active operations at Peenemude during WW II.[1] He was also the person to whom Werner Von Braun then reported. Von Braun was at NASA but came to visit us at Bell often in relation to our involvement in the ongoing NASA Apollo Space Program. At the end of a workday, we would sometimes go to dinner, where the conversation in English would soon transition to discussions of the past in German, a fair amount of which I understood. Listening to such discussions of the past, the contrast and conflict of working for a person who directed a U.S. corporation's research one day, and testified on war crimes the next, was a firsthand study to me in the real futility of war.

---

[1] Dornberger, V-2, 1979.

During these same years, I also began teaching at a local community college in the evenings, while working full-time, motivated by the idea to share the knowledge I was gaining in my industrial position.[2] I never realized until many years later that perhaps this was to be taken as an indication of my true interest, an academic career.

Consequently, my industrial experience in my first position added not only breadth and depth but served as a first step in combining education, engineering, and a global view in a real world setting. I would soon discover though, that such integration would be best pursued as part of an academic career.

THE START OF AN ACADEMIC CAREER - ADVANCED DEGREES

The importance of education was instilled in both my wife and me as small children. Newly married at a very young age, and with me starting a new job, we both nonetheless continued our education; she completing her undergraduate degree as I pursued a Masters degree in engineering. After obtaining my Masters, I continued to take courses to study certain subjects in depth, finally because of interest alone. I soon found out that you cannot just take courses and learn. You need to register in a degree program. Feeling I had all the degrees I would even need, I nonetheless acquiesced and registered for a Ph.D. Soon I found myself taking and passing the doctoral exams to further accommodate the university process, but I continued with courses because finally I found learning fascinating. Learning for the sake of learning was enough for me. Perhaps my parents were right—capability evolves to interest. It was during this time that my interest in education and engineering evolved into an intellectual and emotional passion—one that would soon become my profession as well.

As I continued my education under the umbrella of the doctoral program, I was delighted to learn that the company would give me a year off with pay to complete the university's residency requirement. They were beginning a new program to permit select working professionals to continue their education at company expense, to keep the company current (and me at the same time) in the ever more rapidly changing technology. Later, when I became responsible for education for the working professional (EWP) at Rensselaer, I not only could identify with students who are working professionals, but I could also fully understand and be especially sensitive to their dual roles and responsibilities. My own experiences as a working student made me a better dean.

So partially planned and partially circumstantial, I became Dr. Gerhardt. The Ph.D. made me very attractive to academe and served as the ticket of admission to what became the rest of my professional life.

---

[2]Interestingly, one of our now grown sons, a lawyer, also sought out and began teaching law at nearby colleges. Only time will tell if this becomes as meaningful for him as it was for me. I wonder if this has something to do with that saying about the apple not falling far from the tree.

INTEGRATING ENGINEERING, EDUCATION, AND A GLOBAL EXPERIENCE

I had gained some notoriety for my technical work at Bell through publications, patents, and space program related publicity, and I began receiving offers from the academic world. Because we had some close friends at Rensselaer Polytechnic Institute, I agreed to drive down and be interviewed, in part so we could visit them. Seizing another window of opportunity without great insight or planning on my part, our family of four moved to Clifton Park as Dr. Gerhardt became Professor Gerhardt. Looking back, this was one of my biggest career decisions. Things were going very well for me at Bell. I had risen rapidly, technically and managerially, and was given the Bell Outstanding Management Award. I was in a fine position that appeared to have an excellent future. I chose to try out academe for reasons which I did not fully realize then. I relinquished a solid position with a fine future for a somewhat more uncertain opportunity. At the time, my father thought such a change was unwise, to say the least. I must admit that I initially felt somewhat the same way recently when one of our sons announced he had relinquished a fine position to accept another. As it turns out, for both of us, it was the right choice. Going with your "gut instincts" sometimes really pays off. Even after I made this career change, Bell kept me on as a consultant for several years, and thus created a fallback position that conservative me could not resist.

At Rensselaer, I was soon tenured and promoted to full professor. I moved to academic administration very early in my career when I was selected as the first Chairman of the newly-merged Electrical, Computer, and Systems Engineering Department (ECSE) following a nationwide search. At Rensselaer, my specialty evolved as Digital Signal Processing, emphasizing image processing, speech processing, and brain computer interfacing, work which I continue today. I continued in my professorial role teaching every semester and conducting externally sponsored research, while holding a variety of academic administrative positions including center director, associate dean, dean, and vice provost. I always enjoyed teaching and conducting research as I felt it offered me the chance to reach out and help many students, while the academic administrative roles permitted me to reach out and help many faculty and the university as a whole. For over three decades, I did not relinquish one for the other, and I have no regrets in so doing. The research and teaching at both undergraduate and graduate levels, combined with consulting, helping to form new companies, and the almost daily challenges that new areas such as information technology, biotechnology, and nanotechnology opened have made these years pass all too quickly.

I had a great many professional and personal pleasures during this time, including serving in these administrative positions, winning awards, publishing, and meeting fascinating individuals; however, in the context of this book, the flexibility and opportunities for travel, domestic and international, afforded my family and me are by far one the major advantages and joys of these years in academe.

My technical research and consulting took me all over the world, including most of Europe and Asia. As a result, universities and industry in these areas have turned out to be academic partners and sponsors, respectively, in the international programs I have helped initiate and develop. I also had the privilege to sit on the International Advisory Board of the ASEE and to publish on the subject of globalization of higher education. Each of these experiences added its own dimension of globalism as well, and I relish the opportunities I have had to travel. The consultancies, particularly for international governments, exposed me to many different cultures from the "inside." I was made aware of the difficulties faced by the citizenry of many countries, underdeveloped and developed, and that gave me a much greater appreciation of our good life here in the United States. We have much to do to maintain ourselves and help develop others. It also made me aware how important such exposure is early in life and made me a believer in the importance of globalism and a global perspective for American students in particular. How else can one appreciate that a country like Greece with fewer resources for R&D than MIT, for example, can effectively compete educationally, militarily, and culturally? Travel is broadening yet humbling, educational yet frustrating, rewarding yet challenging. This extensive travel, above all, solidified my passion for seeking to provide all American students with a global experience, beginning with and focusing on engineering.

That same travel, however, provided less than pleasant images that I cannot forget, and in a different but equally important way further fueled my developing passion for global engineering education. These include a child of not quite ten years old, hawking fish on the street with a glazed look, the result of not enough sleep the night before—a child whose formal education had already been completed. Another is a family living in a lean-to of corrugated metal, the back of which was provided by the imported marble of the elegant high rise next to it. This was not TV. This was real. I felt grateful for whom we were and what we had, but that was overshadowed by the realization of my remaining obligation to others less fortunate, especially the children of this world. Why do I care? My family and I were fortunate to be born in a place where you could freely dream, and awake to realize those dreams. I have learned to appreciate that and have been given the opportunity, ability, and position to try to offer that to as many others as possible. This is best done by providing such windows on the world to as many students as possible to instill that motivation in them and leverage them to do the same.

The important point here is that my academic environment and the globalization of engineering education, became part of my life and work and inextricably intertwined, through international research, daily exposure to an international student body, and my consulting activities. My doctoral students are spread worldwide, either in their home countries or elsewhere, and barbecues at our home look like the United Nations. I hasten to also add that each and every global interaction originated because of my role as a Professor and was associated with some aspect of my technical work and expertise. To me, that is important as it speaks directly to the need to consider the S&E workforce as a global resource. We all need to more fully appreciate that the S&E workforce comprises only about 5% of the workforce in the U.S., yet is said to account for about 50% of the U.S. GNP. Personally and most importantly, the globalization of engineering education in which I had become so immersed spilled over to my family providing our sons with much the same everyday diversity I experienced in my own youth that I so cherish today. As such, I am grateful both as an educator and as a father.

My position as a NATO delegate for the Scientific Affairs Division deserves special note as an example of how my professional and personal life became inherently one. We virtually always traveled as a family, taking our two sons along even as babies. Consequently, over an extended period the other delegates became friends, and several like family. We had at least two meetings each year, one in Brussels at NATO, and one in a host country at some unique location. Our children played with other children all over the world and could hardly wait until the next get-together. The celebration they held in honor of my wife's birthday in Athens is a memory that will last forever. Our younger son learned about Greek history in the shadow of the Parthenon on the Acropolis as recounted by the Greek delegate, whose story was then immediately countered by the Turkish delegate. He well recalls their yelling and screaming at each other about the loss of the statue of Athena and the destruction of the roof of the Parthenon, but mostly he remembers them walking away arm-in-arm as friends. He was also emphatically told about the location of the Elgin Marbles by the UK delegate, although discussions ensued about their rightful owner. My wife will not forget the emotional discussion she had in Valencia, Spain with the wife of the Spanish delegate about raising teenage children while I played tennis with her husband. Interestingly, neither spoke the other's language very well, but they came to total agreement as mothers on all issues relating to the next generation. Their tears of understanding were testimony to the blurring of boundaries on our planet. As you cannot see the borders between countries from space, you could not see any barriers of understanding in the language of their tears. To convey that best of all was the banquet that always occurred in the host country. When their traditional local music was played, all the delegates got up to dance, each having learned the same steps in their own country, and thinking their country originated the dance. It was always abundantly clear we all shared and held close the same basic values as people. To this day, we remain close friends with many of those we first met years ago. Several attended our son's wedding, and we visit with these friends almost annually. But no matter how friendly, certain habits cannot change. Our Danish friends regularly stay at our home, and freely use our sauna, but, oddly enough, we always seem to find ourselves too busy at these exact times to join them in their traditional nude practice.

CREATING MY FIRST GLOBAL EXCHANGE PROGRAM

It became clear that professionally and personally I was now fully committed to the concept and importance of globalizing engineering education, and this had become widely known in certain circles of higher education. As a result, I was asked to help co-found the Global Engineering Education Exchange Program (Global EEE) in 1995 and was elected to serve as the President of the U.S. schools and Chair of the Executive Board for the first decade of its life, a role to which I was re-appointed in 2008.

In more detail, in 1994, a group of engineering educators, known to be interested in (or more correctly, passionate about) including or expanding the international component of their engineering educational programs met at the University of Cincinnati to discuss possibilities. The main impetus for this initiative was Dr. Gertrud Humily, a dynamic, motivated, and truly passionate individual on the subject of globalizing engineering education. I had met Gertrud years before in Paris while I was there for a NATO meeting. She was extremely well-connected with universities, industry, and governments in Europe, and sought to develop an opportunity for (originally) European students to study in the U.S. on a reciprocal exchange basis. By the meeting's end, agreement was reached on a proposed structure, initial academic members were committed by people who in some cases could not really act on behalf of their university but did so anyway, and I was elected by the President of the U.S. universities and Chair of the Executive Committee. I was later to be reminded that I had agreed to a subsequent trip to Europe to sign the agreement (which did not at the time even exist) with several European partners. The emphasis in this group was on the product rather than the process. It was years later that specific details such as Bylaws were even drafted. I subsequently did go to Europe where many universities awaited us with open arms, due to Gertrud's associations with nothing less than the rectors of the universities we visited. Gertrud ran the Program from the European end while we in the U.S. engaged the Institute of International Education (IIE) in NYC, in the person of Peggy Blumenthal, IIE Executive Vice President, and her staff to administer the Program. This Program basically was started on TRC (trust, respect, and confidence) in friendships long established, and a desire to get things going quickly by passionate people. It was the very passion of these people, their commitment to purpose, and their quality that made this so appealing to me. In other words, initially it was championed by champions. We had to and did work for years to make this Program self-sufficient and not dependent on any single individual or sponsor.

I was the one who naturally served and still serve as the point person at Rensselaer for Global EEE. Along with my Administrative Assistant, we received applications, did acceptance letters, and received the students when they arrived. All campus services and support did their part as well. IIE did the matchmaking and the rest, all for a modest annual fee from the members. As long as I continued to perform my professorial and administrative responsibilities uninterruptedly, Rensselaer encouraged my involvement, with some financial support for travel expenses and the modest fee we paid to IIE. I maintained an average net zero flow of students and regularly reported that to the President. My past performance in my administrative roles and my close contacts with the Presidents and Deans boded well for my virtually exclusive responsibility for this Program, albeit without any official "international" title. I also helped raise external monies from industrial and governmental organizations to help build support for the Program. For example, I led a multi-year award from FIPSE to do a comparison of the engineering curriculum of seven countries, leveraging on the Global EEE Program, among several other leadership activities.

As formulated, this was to be an exchange program focused on engineering undergraduates, and included computer science whether it was in the engineering school or not. The uniqueness of this consortia-based initiative is that rather than an agreement between only two universities, there is a common agreement all academic partners sign, and exchanges are not required to be balanced directly between any two partners. Rensselaer, for example, could send a student to TU Munchen and receive a student from DTU in Denmark in another year to achieve an overall balance. In other words, it is integrated over time and space. The goal is for each academic partner only to maintain a net flow (students coming and students going) of zero over a period of say three to five years, in terms of student semesters designated as "chits." It is also a voluntary program. It is interesting to realize this was all set up during that first meeting in 1994.

Other advantages include a centralized and long-successful administrative arm (IIE) to process the applications and do the matchmaking to assure this balance is maintained, along with local control of the academic aspects by each of the university partners. This includes the student mentoring and advising, with respect to courses to be taken and their equivalences and how grades or credits are handled. In other words, local academic control coupled with centralized administration has worked wonderfully well for what is now know as the Global Engineering Education Exchange Program. Also, it operates on a tuition swap basis. Tuition continues to be paid to the home university, while room and board is paid to the host university. No tuition is charged to incoming exchange students.

This program has grown and matured and, after fifteen years, handles several hundred exchanges annually. It has about eighty university members from eighteen countries, primarily from Europe and Asia, but from Australia and Mexico as well. There are about thirty-five U.S. universities. Over the years, we have raised support for it from industry (Ford, ATT, ABB) and government (NSF and Department of Education/FIPSE). U.S. members are charged a small annual fee of $2,500. These monies go to IIE to pay for the staffing of the Program. The positive results and success of this Program also demonstrates the enthusiasm U.S. students have for universities that teach in English. Global EEE has approximately fifteen universities that do so, and their immediate draw upon entry to the Program attests to their appeal to U.S. students. Interestingly, this includes some universities in Germany, France, and Spain, which traditionally teach in their native language. There has been a softening of long-held traditions in Europe, especially, to teach undergraduates in the native language while teaching a more diversified graduate student body in English. The Global EEE Annual Conference we established, which is hosted in the U.S. and Europe in alternate years, has proven an extremely valuable mechanism to assure effective networking, provides a regular forum for exchange of new ideas and improvements, and welcomes new members. I recommend this as a requirement for all such global programs.

Finally, it was very important to engage our professional and governmental organizations, most importantly ABET. As ABET was in the process of reformulating their evaluation procedures and process heading towards ABET 2000, we were able to gain their support for such an international initiative by providing proper outcome assessments to evaluate the success of Global EEE as part of an ABET Review. I was invited to speak to the ABET Board about the importance of an international experience early in our process. In some small way, I think this Program contributed to the formation of ABET Inc. and the expansion of their role to offer full accreditation to international universities under the leadership of George Peterson, their Director. Financial support from NSF and FIPSE also opened the door for us to gain their wisdom and help to guide and grow the Program. We also kept these organizations up to date by having them participate in our Annual meetings from time to time, even well beyond the grant period. This speaks to the strength of a consortium as compared to an individual university. As described elsewhere, another advantage is low cost and time required from the home university and the student.[3]

There are many benefits to an undergraduate exchange. Most obviously, an exchange allows universities and students from both "sides" to benefit from a study abroad experience. The students who go abroad benefit from the exchange, as do students in those classrooms that receive them. By hosting international visiting students, home students benefit from an international perspective in their engineering laboratories and classes. In addition, because undergraduate study abroad is typically in the middle of one's degree program, the student and home campus benefit for the remainder of the returned student's time at home. Likewise, the returned international students serve to inform their colleagues on the home campus.

Additionally, the partner universities learned that centralized administration offers many advantages. Global EEE academic partners need only a single agreement with IIE and immediately have an agreement with all other members. Universities require less overhead because they need to maintain only one exchange agreement instead of several separate agreements. A single exchange balance integrated over time and space is maintained, instead of several individual exchange balances. This frees each member institution from needing to maintain balances with each partner campus. Because of this single exchange agreement, ad-hoc exchanges can easily develop between member universities. These include faculty exchanges, team-taught courses, and so forth. The central administration is also beneficial because the tuition swap methodology allows students and universities to have predictability with regards to cost, as students are still considered enrolled at their home institutions. Finally, all academic issues (grade or credit transfer, course equivalencies, and admission standards) are under the control and authority of the individual academic partners.

ISSUES OF SCALE

As with any program, there are some drawbacks to undergraduate exchanges generally, and the Global EEE type of consortium-based exchange more specifically. The exchange is, of course, limited to the partner universities. In addition, as with most international programs, language competence is a factor, which limits the number of eligible participants because of the generally inadequate language skills of U.S. undergraduates. However, as said before, Global EEE has many universities that teach in English to mediate that, and they serve as the greatest draw for U.S. students.

Above all, it is voluntary. This is the most significant drawback, since overall the number of students who seek such an international experience is less than 2% of U.S. students in higher education under any program. As a tuition swap Program, we must increase the number of U.S. students going abroad or, alternatively, limit the incoming international students to maintain a net flow of zero for all U.S. universities. The former is the only realistic option to pursue for the U.S. The lack of participation by U.S. students in international exchanges reflects the core issue of scale-up, in this and other programs. This student semester chit imbalance is further exacerbated since

---

[3] Renganathan and Gerhardt, "Incorporating Global Perspectives in U.S. Engineering Education," 2008.

international students prefer to come to the U.S. for a complete year whereas our students prefer one semester abroad.

This Program was working and continues to work very well, but being voluntary, it always remained a small percentage with respect to the overall U.S. student body. We needed to get the sense of importance out to those students who did not seek out an international experience voluntarily. Thus, I began what was to be several years of advocating the importance of an international experience to our students via their Senate, to our faculty, to our administration, to our President, and to our Trustees. I began with a "platform" of the three numbers: 1.5%, 19%, and 96%. Only 1.5% of American students in higher education engage in an international experience of any sort. Only 19% of Americans have passports. And 96% of humanity lives outside the borders of the United States. This "platform" got everyone's attention, but it still took time to sink in. I finally pointed out that Rensselaer could be first in the country to call for an international experience as a requirement for all engineering undergraduates. The next step was to be REACH.

TRYING TO REACH MORE STUDENTS

In 2007 and 2008, as Director of International Programs for the School of Engineering, I formulated and did the initial development of the Rensselaer Engineering Education Across Cultural Horizons (REACH) Program, based on the basic principles and tenets of Global EEE.[4] The ultimate goal of REACH is a required international experience for ALL undergraduate engineering students. By that time, I had gained the support of Rensselaer's President, Shirley Ann Jackson, and the Board of Trustees, traveled to various international universities to discuss enlisting them as academic partners, gained agreement of the first two academic partners (DTU in Denmark and NTU in Singapore), set up the organizational structure and responsibilities for REACH on campus, hired initial staff, and set up guidelines for future academic partners and program growth. I hope it becomes my legacy at Rensselaer and is used as a model for other U.S. universities.

To some degree, that hope has already become reality. I was invited to participate in a special Summit Meeting held 5-6 November 2008 in Newport R.I., where a group of twenty-three distinguished engineering educators convened to formulate what has become known as the Newport Declaration. It appeared in the February 2009 issue of the ASEE *Prism*, and in an NSF report published in 2009 about the Summit entitled, "Educating Engineers as Global Citizens-A Call to Action." This declaration state the following:

> TO THIS END, we call on engineering educators, engineering administrators, and engineering policy leaders to take deliberate and immediate steps to integrate global education into the engineering curriculum to impact all students, recognizing global competency as one of the highest priorities for their graduates; and TO THIS END, we call on funding agencies, foundations, and leaders in the private sector to shape their

[4]For more information on REACH, see Gerhardt and Smith, "Development of a Required International Experience for Undergraduate Engineering Students," 2008; Gerhardt, Rensselaer Extends its REACH," 2007–2008.

policies and priorities in support of these goals; and furthermore, TO THIS END, we urge that this document be widely distributed and endorsed by all key constituencies.

The REACH initiative was the first of its kind to be established, and the Newport Declaration that followed shortly thereafter is the recognized basis for moving towards requiring an international experience for all engineering students as part of their undergraduate education.

For those interested in replicating such an initiative, and to more fully appreciate how this was accomplished in what might appear to be a very short time and the associated degree of difficulty, it will prove valuable to recount the history that preceded the establishment of REACH.

The fifteen years of experience with Global EEE proved to be extremely valuable for me and for Rensselaer, and certainly for the students who availed themselves of the opportunities afforded them by the Program. At Rensselaer, we gained confidence as we learned how to run such a program. We learned from the experiences of other partner universities, and certainly from the fine personnel at IIE. We were not only able to enhance and refine the operational logistics but to actually change the culture at Rensselaer and develop strong, genuine support for international education from students, faculty, and administration. During that time other international initiatives on campus gained momentum in the School of Architecture and the School of Management, and so did other campus-wide initiatives such as international summer experiences. The benefits to the students in any of these initiatives are by far the greatest reward. As an outcome assessment, consider that every student has one way or another cited their experience abroad as a "life changing event," regardless of where they went or how they performed academically. This phrase speaks to both the narrowness of their experiences to date as well as to the richness of what they have experienced via these programs. As the Chinese proverb says, "It is better to walk 10,000 miles than to read 10,000 pages." In my words (or those of NIKE), Just Do It.

For many years, through my own travel and experiences, I well understood the value of a broadening international experience, and I had created strong linkages with both international universities and industry. Given our experience at Rensselaer, I felt it was timely and appropriate to take the bold and unique step to assure such an experience for each of our students by developing a program that made it a requirement. As I outlined to our students, faculty, administration, advisory boards, President, and Trustees on every available occasion, we would start with our largest school, the School of Engineering, which is about 60% of the university, and ramp it up over a period of years, thereafter extending the concept to our entire campus.

Why timely? Rensselaer's President Shirley Ann Jackson had taken many bold and unique steps since coming to Rensselaer. These included major new research initiatives, such as creating a Center for Biotechnology and Interdisciplinary Studies, a new Biotechnology building, an Experimental Media Performing Arts Center building in association with another major initiative, and several others. Under her leadership, the terms "global reach" and "global impact" became part of our vision and mission for the university as part of the overarching Rensselaer Plan. But more than that, our President stood above all for diversity in all respects, and she sincerely believed in a global reach and global impact for Rensselaer as a dimension of the diversity she sought. Additionally, we had just hired a new Dean of Engineering who was likewise committed to globalizing engineering education.

Finally, with their enthusiastic support, I began in the leadership role of REACH. And for the first time in all the years I had been leading the Engineering School's international efforts, I was given the title of Director of International Programs. As a Professor in two departments and with other academic administrative responsibility and titles, I did not need or seek this title. However, it was a first and clear recognition that Rensselaer had arrived at the point of fully and formally acknowledging the importance of this position. For that alone, I was appreciative. My own personal long-term vision was coming to fruition. By all measures, the time was as right as it could ever be. And so we all began.

REACH formally began in spring 2009 with the junior class. Initially, about 20% (50 students) of the then junior class participated in a study abroad semester at a partner university, and Rensselaer reciprocated by hosting an equal number of the partner university's students. It is anticipated that a level of 25% of engineering juniors will be maintained for two years, increasing to 50% for two years, 75% for two years, and then reaching 100%. At that time, if not before, campus-wide participation will be addressed. With the size of the junior engineering class at about 600-650, twelve to fifteen partner universities will be needed. The initial partners will be in Europe and Asia, and expansion to other continents will follow. Although the major component of the mandatory program will be a semester study abroad experience, other options may include the Semester @ Sea and special summer initiatives among others. Students who need to opt out could participate in an on campus international component of the program, utilizing the university's base of international graduate students.

During the developmental stage, one needs to handle the day-to-day operations via an Institute-wide International Program Committee to address the major topics of program development and implementation. I created five sub-committees to handle academics, housing, student services, health and safety, and administration. These can and do address both small and large issues. Early on, I had to engage the student body (via the student Senate), faculty, and administration; develop an Academic Partnership Agreement and get it approved by legal counsel; allocate time for formulating academic templates for each of our curricula in cooperation with each of our initial academic partners; and address the issue of staff and begin hiring.

My off-campus activities were mainly focused on recruiting academic partners. Here again, our extensive prior experience with Global EEE proved to be invaluable. A long history with NTU and DTU, and the associated trust, respect, and confidence (TRC) established over that time, made it a relatively easy transition to gain their agreement to join REACH as founding partners. The contacts were, once again, at the Presidential level, contacts I had developed over many years. To minimize potential challenges in transition to this large scale-up, I sought academic partners of the highest quality with a diversified portfolio of curricula that matched ours and with a large existing base of international students so that our fifty students would not burden them. We also sought universities that taught many course offerings in English so as not to further limit our U.S. students. Although

I strongly support a total immersion concept in principle, pragmatically it presents many difficulties that needed to be avoided for a quick and successful startup. Finally, we selected universities in desirable and safe locations. Our intent was to get a large number of exchanges under our belt before we expanded to less-developed countries, which is our plan for the future. We will also seek to match technology needs of developing countries with the student mix selected for that country. Three other universities are "waiting in the wings" to join REACH as I write.

Endorsed and supported by Shirley Ann Jackson, REACH was officially launched on 11 April 2008 at a special Presidential Colloquy organized for that purpose. Both the Presidents of DTU and NTU, our first academic partners with whom I had worked for more than two decades, participated. Other colleagues from DTU and NTU, the Vice President of the IIE in Washington D.C., and one of my former students, an entrepreneur extraordinaire and global citizen who was at the time helping to rebuild housing in Iraq, were also in attendance. All these had become not only colleagues but friends. On that background and foundation, they all readily accepted President Jackson's formal invitation to help initiate this first step in establishing a requirement for an international experience to be made part of the engineering curriculum. Our own students actively participated in these activities, which were widely publicized on our campus and later on the campuses of our academic partners with special signing ceremonies. With this kickoff, the REACH program received the notoriety, Presidential support, and visibility it needed and deserved, both on and off campus. It also received high-level attention in our fund raising campaign as well.

The REACH Program is a School-Centric initiative, and is different in some ways from the Global EEE Consortium-Centric Program, yet similar in others. Rensselaer plans to handle the development, implementation, and logistics of REACH, as compared to outsourcing this role, as the IIE handled the administrative aspects of Global EEE. Whereas Global EEE is voluntary, REACH will become mandatory. Whereas Global EEE permits taking courses at the host university subject only to the approval of the student's academic advisor, REACH has structured course templates for outgoing and incoming students by curricula and partner. Under REACH, Rensselaer determines the mix of students going to our academic partners (who goes where) rather than leave it to the student to decide. We initially chose to transfer grades under the Global EEE program, but will transfer credit under REACH. Such guidelines or constraints become necessary when mandating a requirement and not extending the time to degree. Finally, whereas Global EEE does now scale up easily, REACH is quite scalable with the addition of more academic partners.

University-Centric programs are campus-oriented solutions that allow the maximum in targeting student audience, and create immediate benefit to the school. As university-centric models, the schools immediately benefit from having a globally-competent class of students in the sciences and engineering who are immediately attractive to graduate institutions and industry. In addition, the curriculum is university-created, so the faculty does not view the international experience as an interruption in the academic trajectory. The main drawbacks of such university-centric programs are that they require greater cost to the institution or student, and a substantial investment of time by faculty and staff. This needs to be recognized not as a burden, but a responsibility, for the individual and for the university.

LESSONS LEARNED DURING A LIFE OF GLOBAL EXPOSURE, ENGINEERING, AND EDUCATION

Personally, I have learned to appreciate my own life and the windows of opportunity that opened along the way. Being born in the United States at a wonderful time in its development, I have been able to participate in that development in some small measure, for which I am grateful. I am grateful to have shared my life with an exceptional and understanding family, now including two grandsons. I have known only freedom, I have never known hunger, and I have always slept in a clean bed. I wish that for all citizens of this planet.

Professionally speaking, I offer some advice to those first embarking on a path of globalization of education.

It is most important to follow your passion.

It is important to have trust, respect, and confidence (TRC) in those you work with or for, and in those who you hire. Seek out those who provide that to you in reciprocation. Also remember that in any culture, TRC take time to develop, and little replaces time in the long run. Time and money are not fungible.

Have pride in your past, and faith in your future. However, do not live in the past and do not adopt hope as a singular strategy to assure that future.

Recognize that new initiatives are invariably developed by a "champion" or group of "champions." However, once established, it is critical to ensure the sustainability of the program by cultivating future leadership. Sustainability needs to be defined in terms of not only resources but people. I strongly encourage faculty leadership with proper staff support for an initiative of magnitude in globalizing engineering education. The faculty member ideally should be a Full Professor and carry a credible academic administrative title. The person needs to be respected by the faculty, be an outstanding and charismatic leader as well as an excellent manager, possess fine interpersonal skills, have a university view, and understand the desire and need for fundraising. Because such initiatives must be interdisciplinary as well as global to be successful, they must evolve to being campus-wide. As such, I believe the level and breadth of the administrative responsibility should ideally be a Vice Provost for International Programs.

Finally, to those that still question why global engineering education is necessary, I offer this.

It is because every leader in history always recognized that the future of the country lies with the proper education of its youth and that education today mandates a global perspective.

It is because 96% of humanity lives outside the United States, and hardly 20% of our citizens have passports, and less than 2% of our students in higher education participate in an international experience.

It is because there exists a fundamental interdependence of all countries sharing this planet, including its energy resources, its air, its water; and because a catastrophe for one of us becomes a catastrophe for all of us; and because global issues call for global solutions.

It is because developed countries need to more strongly feel an obligation and responsibility to those less developed, because famine of one country's children is famine for all of our children; and because we still need to learn to prize long-term cooperation more than winning a short-term conflict.

It is simply the right thing to do.

ACKNOWLEDGEMENTS

I extend my thanks and appreciation to all the unsung heroes of all global engineering education initiatives, including faculty, staff, and administrators. Most of all, I salute the students who participated and those who will participate in a global experience as part of their education. May you become all you are capable of being and may your global experience serve to bring you everything you desire and deserve.

REFERENCES

Bhandari, Rajika and Patricia Chow, eds. "Open Doors 2008-Report on International Educational Exchange." New York: Institute of International Education, 2008.

Dornberger, Walter. *V-2*. New York: Bantam Books, 1979. 203

Gerhardt, Lester A. and Richard N. Smith. "Development of a Required International Experience for Undergraduate Engineering Students." Paper presented at the IEEE/ASEE 38th Annual Frontiers in Education Conference, Saratoga Springs, NY, United States, 2008. DOI: 10.1109/FIE.2008.4720335 211

Gerhardt, Lester A. "Rensselaer Extends its REACH." *Rensselaer Engineering* 10, no. 2 (Special Issue Fall/Winter 2007–08). 211

The National Academy of Sciences. "Rising Above the Gathering Storm: Energizing and Employing America for Brighter Economic Future." Washington DC: The National Academy Press, 2005.

Renganathan, Vijay and Lester A. Gerhardt. "Incorporating Global Perspectives in U.S. Engineering Education." Paper presented at the American Society of Engineering Education Annual Conference, Pittsburgh, PA, United States, 2008. 210

Developing Global Awareness in a College of Engineering

Alan Parkinson

INTRODUCTION

This is a narrative account of my experience developing global awareness in the Ira A. Fulton College of Engineering and Technology at Brigham Young University (BYU). The narrative begins by establishing my personal background. I then discuss the context of the institution and students, both of which are important elements in this story. Within these contexts, I explain how I became involved in developing a college strategy that incorporates global awareness as one of five strategic outcomes. I discuss our efforts of the past two years, which have involved sending students to China, Europe, Mexico, Peru, Tonga, and Romania. I conclude with observations about what we, the college leadership, have learned, the challenges we have faced, and what we yet hope to accomplish.

A PERSONAL JOURNEY FOR GLOBAL AWARENESS

I grew up during the space race to land a man on the moon. I was eight years old when Alan Shepherd became the first American to travel into space (on my birthday, no less) and sixteen when Neil Armstrong took the first steps on the moon. I could imagine nothing more exciting or challenging than building rockets to go to the moon. The Saturn V rocket, along with the command module and lunar module, were masterpieces of engineering achievement. The undertaking was bold in its execution and broad in its sweep, involving thousands of engineers working for hundreds of companies. I consider the voyages to the moon to be among the greatest voyages of discovery in human history. Even today, I have in my office photos of the Apollo spaceflights. The moon landings were my inspiration to become an engineer.

After graduating from high school, I was admitted to MIT but chose to attend BYU. BYU is one of the largest private universities in the United States, with approximately 33,000 students, who come from all fifty states and more than 100 countries. It is a religious institution, sponsored by the Church of Jesus Christ of Latter-day Saints, hereafter referred to as the LDS Church. The focus of the university is primarily on undergraduate education. I felt I could establish a solid educational foundation at BYU and then go off to a better known engineering school for graduate work.

I also wanted to prepare my freshman year to go on a mission. The LDS Church has a far-reaching missionary program, with more than 50,000 young adults, age nineteen to twenty-one for men, twenty-one to twenty-three for women, serving at any one time in hundreds of missions across the globe. Such service is voluntary and self-funded. You agree to go wherever you are called. I had been planning on serving as a missionary for a number of years. After finishing my freshman year, I submitted my application for missionary service.

To my complete surprise I was called to serve in Japan. Serving in such an "exotic" place had never entered my mind, and I was somewhat concerned whether I could learn Japanese. Learning languages is not a natural strength for me—I was always better in math than in English. All I could do was commit to work hard. As was the procedure at that time, I spent two months at the Language Training Center in Hawaii for missionaries called to serve in Japan, Korea, and Taiwan. There, from eight in the morning to ten at night, I studied Japanese. I do not recall ever being more mentally exhausted than I was at the end of those days.

After two months studying the language, I continued my journey west from Hawaii with a group of forty other missionaries to the Japan Central mission, with headquarters in Kobe, Japan. The first night we were there, as something of an initiation, all the new missionaries were taken to the public bath (fortunately sex-segregated, although we did not know that until we arrived). As I struggled during the first six months to adapt to everything that was new and different, I felt almost as if I had been transported to another planet. I found people could usually understand me when I spoke, but I could not understand them at all because they spoke so fast. Only after weeks of concentration did my ear gradually become accustomed to hearing where one word ended and another began. As missionaries we spent many hours teaching, contacting people on the street to see if they had any interest in religion, and tracting door to door. It was not unusual to converse with dozens of people during the course of a day. It was hard work, very demanding physically and emotionally.

You feel almost as if you live a "mini-life" during the two years of a mission. You start out as a child, very dependent on your companion; gradually you become an adolescent, able to do a lot on your own; after six to eight months, you mature into an adult, becoming the senior rather than the junior companion in the twosome; and finally you become an old man, training others, before you "die" and go home.

As might be imagined, such an experience is very broadening for a nineteen to twenty-one year old. Missionaries try to live pretty much as the native people do. Often, in foreign countries, a U.S. missionary will be paired with a native as a companion. As expected, this accelerates learning the language and customs of the abroad country. Missionaries are taught to embrace the culture and people among whom they work. Many are profoundly affected by their missions, and many develop a deep love for the country in which they served. I returned home with great respect for the culture and accomplishments of the Japanese people.

I re-entered my studies in Mechanical Engineering at BYU. At first I tried to keep my language skills current by taking classes and associating with Japanese students in the U.S., but after a year,

I felt engineering was about all I could handle, so I let my language skills slide. I graduated from BYU in 1977 and continued on to graduate school at the University of Illinois. My M.S. degree had a focus in materials science, but I switched areas for my Ph.D. and did research in applying optimization methods to engineering design. I attended graduate school with the hope that someday I might return and teach at BYU. The opportunity unexpectedly presented itself immediately upon graduation, and, not sure if fate would present this chance again, I decided to take it.

From 1982 to 1994, I found some success in teaching and research (continuing my work in optimization and design) and advanced from assistant to associate to full professor. At about this time the chair of the Mechanical Engineering department at BYU decided to leave the university. I was the logical choice to succeed him, as I was a full professor (although new) and had always taken an interest in administration. I was appointed chair in 1995.

My time as chair was one of both professional satisfaction and disappointment. On the positive side, we (the faculty) were able to define a strong overall vision for the department. Our mission statement was "to provide a world class education to undergraduate and graduate students in an atmosphere enlightened by the principles of the gospel." I realize the term "world class" has become cliché, but at the time this was not the case, and it represented a new benchmark for us. Relative to any number of criteria, we found it useful to ask ourselves the question, "Are we as good as the best departments we know of?" In some areas of the department, such as our laboratories, it was very clear we were *not* world class. Thus, we became aware of areas we needed to improve. As a side note, in referring to an "atmosphere enlightened by the principles of the gospel," the mission statement references BYU's religious underpinnings and the ideal of pursuing a balance of secular and spiritual knowledge "by study and by faith."

Partly as a result of understanding where we wanted to go, we saw an increase in all areas of resources: budgets, space, and faculty. We renovated nearly every laboratory in the department. Class sizes and faculty loading were reduced. Teacher evaluations, already at good levels, continued to climb, and research publications increased. We strengthened a nationally recognized program in capstone design, and we started a new joint MS/MBA program with the business school.

This was also during the time frame of ABET 2000, which acted as a strong driver for change in engineering education. I was very much in favor of both expanding the set of desired attributes for engineers and moving from a prescriptive accreditation approach to an outcomes-based approach. However, the path to achieving ABET 2000 objectives was new and unknown, and the faculty spent a fair amount of energy figuring out how to proceed. And, as is still the case, some faculty members were skeptical that the changes were worth the cost or would lead to lasting improvement. I recall at that time I had a quote from Machiavelli on my bulletin board that said something to the effect, "there is nothing more difficult or uncertain than to initiate a new order of things."

On the negative side, the college leadership asked us to absorb part of a department that was being disbanded. The faculty coming into ME wanted to continue their program as a separate program within the department, which was resisted by many ME faculty. As might be guessed, this led to a lot of discord. Also, because research funding for the department per faculty member lagged behind that of our sister departments, we came under a fair amount of pressure from the college to improve. Eventually I put our ABET 2000 efforts on hold and turned our entire focus to improving scholarship and research funding. I came to understand that it is very important to pay attention to the agenda of those over you. And I felt the sting of criticism from faculty who felt things should have been done differently.

I often pondered during this time how to best effect change in an organization as seemingly complex as a department, where change must occur through consensus building and group effort. I came to several conclusions, which are still being evaluated in the even more complex environment of a college with eleven programs. First, I came to feel that it was better to focus on a few key areas and make measurable progress in those areas than to try to move ahead on a very broad front. Second, there comes a point where it is better to make a decision and move ahead than to continue to discuss and debate a course of action. Third, in the context of an organization which still has teaching and research to attend to, change is accommodated better if done at a slow and sustained pace, rather than through a huge one-time effort.

My interest in organizations led to a somewhat unusual diversion in 2001. Having finished my term as chair, I decided to enroll in the executive MBA program at BYU. I had often wanted to learn what they teach at the business school. I have always had an entrepreneurial bent. While an assistant professor, two other faculty members and I developed some optimization software and formed an off-campus company which sold the software to companies and universities in the U.S. and Europe. Helping to run this little firm was fun for me. Although sometimes not recognized, there is a lot of overlap between business and engineering. Most engineers' salaries are paid for, ultimately, by services or products which are purchased in the marketplace. I felt I could benefit from understanding better the business context in which engineers worked. Although going back to school at age forty-eight was somewhat outside my comfort zone, I thought this would be good for me. It could hardly have been more convenient since I worked at the university. So, for two years I went to school at night and during the summer to complete my MBA degree. It was a little strange to have homework, take tests, write papers, and receive grades, sometimes from colleagues I worked with on university-level committees. On the whole, however, it was a very enriching experience that deepened my interest in and appreciation for areas such as strategy, organizational change, human resource management, and leadership.

At the end of the executive MBA program, it is usual to take a two week trip with students to either Europe or Asia. For my MBA cohort, a trip to South America was substituted in place of Asia because of the SARS epidemic. This trip was also, like my mission, a broadening experience. One overriding impression was of poverty and the struggle of providing the basic necessities of life. In many of the countries we visited, the cost of living was not that much lower than the U.S., but wages were much lower. In terms of engineering, it was apparent that it was difficult to compete with more developed countries when the physical infrastructure (communication, transportation, manufacturing facilities, etc.) was not in place or when a cadre of well trained engineers was not available.

Based on my activities as chair and my interest in areas which extended beyond traditional research, I was asked to be associate dean in charge of undergraduate programs in 2003. I was very interested in understanding the changes happening in the world and their implications for engineering education. I sensed some of the observations and conclusions later made by Thomas Friedman in his book *The World is Flat*,[1] even if I could not articulate them. At one point I presented to the faculty the idea that the rise of skilled, inexpensive engineering workforces in China and India could act as a "disruptive innovation"[2] relative to engineering in the United States. U.S. engineers therefore needed to add additional value, beyond technical skills, to remain competitive in the worldwide marketplace. Recalling my MBA trip to South America, I decided with a colleague in Mechanical Engineering, Spencer Magleby, to organize a similar trip to Japan, China, and Hong Kong for engineers. Fortunately, as associate dean, I had a lot of latitude in deciding to do something like this as well as some resources to draw on. We started from a knowledge base of just about zero. For months I spent about half my time organizing this trip. Since a lot of this type of trip is arranged through contacts, and I did not have contacts in Asia, I had to develop these. Typically, I would do this by finding someone who did know someone in China and who could provide an introduction for me. Developing these relationships was time consuming and sometimes stressful. As an example of the pressure I felt, I did not attend the opening picnic with the students because I was on the phone trying to nail down visits in Japan, just a few days before we were to leave.

As we got ready to go, Spencer and I wondered if we were a little crazy. What if we arrived in the middle of China and no one was there to meet us? What if someone got sick or got lost? What if my thirty-year-old Japanese couldn't get us out of a tough spot? We felt keenly our responsibility for the group. We needn't have worried: the trip turned out to be "charmed." Everything went well. We had outstanding visits with companies and engineers in Japan, China (visiting several cities), and Hong Kong. Even though I had anticipated what we would see, being there and seeing the scale and pace of change in China made a deep impression on me and on the students. I vividly recall visiting an electronics toy factory where the head engineer described how they could go from a customer's concept to a product on the shelf in six months.

One of a number of incidents became a metaphor for me. While driving down a freeway in China, I noticed that the billboards advertised the manufacturing capability of various companies (ISO 9001 certified!). I reflected that in the U.S. the billboards would have been for lawyers. A student made the same observation about their phone books: the front pages carried ads for manufacturing companies instead of for attorneys. I felt these observations were representative of the difference between the two countries: China is building its economy by building things, and our economy is more and more built upon services, some of which I do not feel create real wealth for a nation.

It is not an exaggeration to say the participating students experienced a huge paradigm shift. Previously they had considered (even if they were too polite to say so) that China was a place for cheap labor and technologically unsophisticated products. They came away feeling the Chinese could

---

[1] Friedman, The World is Flat, 2005.
[2] Christensen, The Innovator's Dilemma, 1997.

manufacture with the best of any country and were gaining ground fast in all areas of engineering. They sensed the spirit of entrepreneurship that pervades the country. The pace of infrastructure development was astounding. From the China headquarters of Cummins Inc., in Beijing, we counted dozens of construction cranes. As a colleague said to me, "There is a train on a track coming right at us. If we don't move we will get run over."

In 2005 the term of the current dean came to an end. I had served as associate dean for two years. I was asked to apply. Not anxious to be in the hot seat again, it was with some hesitancy that I did so. I did feel, as I did when chair, that the college should move ahead more aggressively in defining a strategy to address the challenges confronting engineering and technology. As part of the application process, I was asked to submit a paper describing past leadership experiences and explaining my vision for the college. I tried to be candid. I spoke of what I felt the successes and failures were while I was chair. I discussed a vision for the college that included, as one element, leveraging off of our students' strengths in languages to develop global awareness. Initially I felt my appointment was unlikely, so I did not worry about it. I had some sleepless nights when I discovered I was one of two final candidates. When the appointment was extended to me, I accepted with some trepidation.

BYU STUDENTS

In order to explain the strategy for the college, which was developed with my colleagues, I need to provide a little more information about the university. BYU has a relatively unique student body. The freshman class of 2008 has an average ACT score of just over 28, which corresponds to the 92nd percentile. By the time they are juniors or seniors, most of the men and about one fourth of the women have served as missionaries. As a result of missionary service, more than 75% speak a second language.

In addition to academic qualifications and relatively unique cultural experiences, another attribute of the student body is leadership potential. For example, in a recent freshman class, 28% of the students were high school student body officers or team captains (and yes, I am surprised the Admissions office tracks this data). As part of a mission, a seasoned missionary might be responsible for directing the efforts of a district (eight to ten missionaries) or a zone (twenty to fifty missionaries). LDS youth also have numerous leadership and mentoring opportunities while growing up. Groups of young men and women take turns serving in leadership roles among their peers in age groups twelve to fourteen, fourteen to sixteen, and sixteen to eighteen.

Reflective of another dimension of the student body, students at BYU commit to live an honor code, which stresses honesty and integrity. The honor code is typically not something which exists external to the students' values but is rooted in deeply held religious beliefs. I do not wish to imply that BYU students are unusual or unique in aspiring to these character traits, but this overt pledge does form a strong foundation upon which to build a commitment to professional ethics.

DEVELOPING A COLLEGE STRATEGY

BYU represents an enormous investment by the LDS Church, which covers about 70% of the operating expenses—an investment of hundreds of millions of dollars per year. In consideration of this investment, and in consideration of the significant potential of the student body, we asked ourselves questions such as, "How do we become better than we are now?", "Can we be a leader among engineering colleges, and, if so, how?", and "How do we prepare our students to succeed in an increasingly global, competitive engineering environment?"

Many engineering colleges have decided that a path to excellence lies in increasing their research activities and growing their graduate programs. There is no question that numerous benefits and prestige accrue with increasing levels of research. For BYU, however, the path must be different. The Board of Trustees for the university has determined that BYU should remain primarily an undergraduate institution. Currently the graduate program is about 10% of the size of the undergraduate program. Although this might grow some, it will not grow a lot. The path to excellence for BYU will not be to become a major research university. It must focus primarily on being an excellent undergraduate institution.

If that is the case, how might this path be defined? Concurrent with our deliberations of a college strategy, the National Academy of Engineering released the document, "The Engineer of 2020,"[3] which discusses the forces acting on engineering in the United States and what preparation engineers will need to be competitive in the global economy. This report has been followed by a number of other publications that support its conclusions. The report indicates that the skill set for engineers needs to expand beyond analysis and technical ability to include communication and leadership skills, creativity, adaptability, ethical responsibility, and a commitment to lifelong learning.

My associate deans and I felt that this set of desirable skills overlaps well with the attributes of BYU students. Given the particular skill set and experiences that BYU students bring to the table, the strategy we developed was to play to our strengths by developing students with the characteristics of technical excellence, leadership, global awareness, character development, i.e., a strong sense of professional ethics), and innovation. To help remember these attributes, we formed an acronym called LIGHT (Leadership, Innovation, Global awareness, cHaracter development, Technical excellence). We wish to leverage off of the life experiences and preparation our students have to develop a strong combination of attributes in these areas. By so doing we hope to position our students to succeed, to provide what the workforce wants, and to be an influence for good in the world. The goal is for the college to be a leader, and to be excellent, in part, by the type of graduates we produce.

Global awareness forms an important element of our strategy. The term "global awareness" is quite broad, and we spent some time defining what it means. Frankly, our understanding has evolved as we have gained experience and evaluated various programs. But from the beginning, we felt one way to acquire this attribute was to run study abroad programs that had a technical focus. We felt there would probably be a number of different ways to do this, and, in fact, we have so far supported several program formats, including what we refer to as mentored travel programs,

---

[3] National Academy of Engineering, *The Engineer of 2020*, 2004.

service learning programs, international design programs, and supervised internships. We have a set of desired outcomes for each of these.

For many years BYU has run extensive study abroad programs, with permanent facilities in London and Jerusalem. Partly because of missionary work, BYU teaches some sixty languages. Yet, perhaps surprisingly, the college had never explicitly tried to leverage off of this unique aspect of its students or this strength of the university as part of a strategic plan. Frankly, given the direction of world affairs, doing so just seemed to make sense to the college leadership. In terms an MBA student would use, this was an undeveloped competitive advantage.

Our efforts to develop technical study abroad programs began in earnest several years ago. Because a number of engineering colleges were already active in this area, it made sense to study what other universities were doing. As part of my own scholarship, I decided to study and then write a paper on engineering study abroad programs. Deciding to do this was perhaps a reflection of my previous experience in teaching and doing research in design optimization. Part of a good design process involves understanding customer specifications, benchmarking similar products, and learning what constitutes "best in class." In design we teach students to avoid the "not invented here" syndrome, where designers refuse to consider using the best features of designs from other companies because they were "not invented here." So as a step in "designing" our first programs, I reviewed programs at about twenty-five universities, often visiting at length via email or on the phone with the program directors. I identified nine different formats, and tried to infer both challenges and strengths of the various formats as well as best practices.[4]

BYU PROGRAMS

With study abroad, there is really no substitute for experience. So in 2007, my colleagues in the college office and I decided to get our feet wet and learn by doing. In our discussions regarding college strategy, my associate deans and I felt we should try to have some sort of presence in China, Europe, and India. These locations were chosen because a lot is happening relative to business and engineering. We felt it would be ideal if we could run programs in these locations on a regular basis that were open to any student in the college. On top of these programs we could have other, discipline specific programs; these would be designed and run more by departments. Also, some interest had previously been expressed by students and faculty in starting a chapter of Engineers Without Borders (EWB). We decided to move ahead with EWB, a program in China, and some department specific programs. Later we could expand to Europe and India.

A STRUCTURED PROGRAM IN CHINA

I had a colleague in Electrical Engineering, Brent Nelson, who had been a missionary in Taiwan many years ago and spoke Mandarin. We live in the same neighborhood and have been on camping

[4]Parkinson, "Study Abroad Programs," 2007.

trips together. I talked to Brent about starting something in China. He was enthusiastic, and his chair was supportive. Of course, it helps when it is the dean asking.

We were not really sure how the program should be structured. Should students just visit companies and universities in China during a relatively short (e.g., two to three week) time period? Should students take engineering classes, in English, at a Chinese university, as some universities do? Brent and a colleague from Electrical Engineering decided to take a scouting trip to observe the possibilities first hand. During this visit he determined that a campus with a well developed infrastructure to host foreign students was more important than one with a highly rated engineering program. As such, he recommended we use Nanjing University, a top tier Chinese university with extensive experience hosting foreign students, as home base. Nanjing could arrange housing for the students and provide instructors and classrooms for courses in Business Chinese or Chinese culture. To supplement these courses, Brent created a course called "Globalization, Engineering, and Technology" and obtained approvals for this course to count as a technical elective in every major in the college. In addition, he worked with the Chinese department on campus so that the language/culture courses taught while abroad would fulfill the requirements for "Chinese Culture" or "Business Chinese." Thus, students would receive six credits, three of which would count as technical electives. The program included six weeks of instruction Monday thru Thursday mornings, interspersed with trips to companies and cultural sites during the three-day weekends.

Why would Brent, a faculty member with an active research program, take the time to create and run such a program? Brent states,

> Prior to going I convinced myself this would provide a unique experience and significant value to the students. After returning I am now certain it did. Importantly, I was a witness as the students in the group began to put their educational experience and anticipated careers into a context they never imagined existed – a global context. It was evident in the questions they asked and their final term project presentations that they had learned to view the practice of engineering in a very different way.

> On a personal level, it was a phenomenal experience to take my family of six to China for two months. They benefited as much as students did, and now ask regularly to make sure that the program will be repeated in coming years so they can return.

> Personally, being involved has been invigorating for me. I get to interact with a diverse set of students from the college in a variety of majors. I feel like, for those who participate in the program, their career view (and often career goals) are fundamentally changed by the experience. Finally, it is building bridges with engineers and academic colleagues in China which I believe will pay off in myriad ways in the future.

I recall getting regular updates via email while the students were in China. It was exciting for me to hear what they were doing. I felt this was a strong program because students had a structured, challenging classroom experience in China that was reinforced with out-of-classroom visits, including several two to three day trips, which allowed students to experience issues firsthand. We have now run this program several times.

LEVERAGING OFF OF GEORGIA TECH IN FRANCE

Somewhat unexpectedly, we had an opportunity to run a program in Europe. This was in partnership with Georgia Institute of Technology, which operates a permanent facility in Lorraine, France. During the summer program, GT-France offers about twenty-five courses, including a variety of sophomore and junior level engineering courses, as well as courses in the French language. Most courses are taught by GT faculty in English. A BYU alumnus, Paul Voss, is a permanent faculty member there. Interested in perhaps recruiting BYU students as graduate students, the administration at GT-France graciously invited BYU to send a group of undergraduates there for the summer.

I jumped at the chance. We wanted to start a program in Europe but had no infrastructure in place. With this invitation, we could take advantage of the infrastructure already developed by Georgia Tech and get something going. The program had two main elements: class work and touring. From Monday thru Thursday, students took engineering courses; the rest of the week was left open for students to travel around Europe. To complement this travel, BYU arranged for several company visits, including an outstanding visit to Airbus.

This was a good learning experience for us as well. Two of our faculty had the chance to visit GT-France while the students were there to learn about the operation of the program. Although students had an enriching experience touring Europe (while taking required engineering courses), their understanding of globalization issues, one of our main objectives, was weak. Indeed, some of the students viewed the company visits as interfering with their personal travel rather than as an opportunity to better understand globalization issues. As a result of these observations, we decided to run our own program in Europe in 2008.

MODELING WATER RESOURCES IN MEXICO

In 2004, the Department of Civil Engineering initiated its own study abroad program in Mexico. The college's only decision relative to this program was whether we wished to support it. We not only wanted to continue it but to strengthen it. It is a discipline-specific program tied to a Civil Engineering course on hydrologic modeling.

At the beginning of the school year, professors at BYU and partner universities in Mexico select a number of actual water projects that could be modeled using software developed at BYU. Possible projects involve analyzing and making recommendations regarding water runoff, ground infiltration, flood control, and reservoir management. Student teams are formed for each project at both places; they work together via email and the Internet. Before traveling, the teams define the problem, determine objectives, and perform preliminary modeling. Most of the communication is conducted in Spanish. Towards the end of the semester, BYU students travel to Mexico for two weeks and meet with their Mexican counterparts. They visit the sites, meet with stakeholders, refine their models, and make final recommendations. Project presentations are given in Spanish. In 2007, seventeen students participated. During the years the program has been in operation a number of the students' recommendations have been implemented.

The director, Jim Nelson, speaks about the Mexico program as a "journey of faith." The genesis of the program can be found in his own missionary experience when he was first exposed to extreme poverty. That experience motivated him, "I felt that if possible one day I wanted to be able to make a difference. I find that many of our students share this same feeling, and while I don't really like to think of this program as service-oriented, there is definitely a component of it that is satisfying for those who participate." However, after running the Mexico program for a couple of years, he wondered if he should continue. Even though he felt the experience was very valuable for students, it was a lot of work, and he suspected some in his department viewed it as a "boondoggle." Jim indicates,

> I was thinking that I probably should not do this anymore, or would at least take a break for a while. At the next fall university conference you had just been made dean. Together with [associate deans] John and Spencer you presented a vision for the college that espoused most of the same things I was seeing students experience as part of these trips: language, cultural awareness and context for engineering practice, global skills, teamwork, and leadership—they were already telling me in their final reports that this is why they valued these experiences. For me it was like, "Yes, somebody else out there does see value in all of this." I had talked to [a colleague] before about joining me because he speaks Spanish and he is inclined towards this kind of thing, but I can guarantee that he would never have tried it had it not been for the vision and support of the college.

CREATING BIO-DIESEL FUEL IN TONGA

About this time, students and faculty, primarily from Chemical Engineering, made a proposal to establish a chapter of Engineers Without Borders.[5] At the college level, we mostly had to provide support and get out of the way.

For the inaugural project, students developed and implemented a small-scale facility for the conversion of coconut oil to bio-diesel fuel on the Pacific island of Tonga. To prepare for this project, students enrolled in a new course titled, "Global Projects in Engineering and Technology." The course was open to all disciplines and counted as a technical elective. As part of the course, students worked in multi-disciplinary teams on one of six elements of the project. They built, tested, and refined the hardware for the project.

At the end of the semester, twenty-four students traveled to Tonga to set up a pilot bio-diesel facility. I was concerned when I heard much of the supplies and hardware they had pre-shipped to the island had not arrived when they got there. Getting the supplies to Tonga in time for them to be used was itself a lesson in cultural differences. Finally, the supplies arrived, and students were able to demonstrate the bio-diesel process to local farmers, as well as to government officials, and salvage the project. I felt we learned some good lessons, which would stand us in good stead next time.

[5]http://www.ewb-usa.org

This type of program fits well with the aims of the university. At the entrance of the university is a sign with a motto, "Enter to learn, go forth to serve." Although educational objectives must remain paramount, combining strong learning experiences with humanitarian activities is something that makes sense to me.

I have come to appreciate the complexity and challenge behind this type of program. First, we have students coming together with a wide variety of backgrounds. How should instruction be designed given the different preparation levels of the students? Second, the students are working on design projects, which tend to be "messy" anyway, but in this case are more difficult because the customers are typically thousands of miles away. Third, the students travel and build their designs at a remote location, hoping they have brought everything they need. Taken together, all of these challenging circumstances take a lot of energy to overcome. I am somewhat concerned whether the faculty directors can sustain their efforts here. We are also learning a lot about sustainability in these types of projects.[6]

COORDINATING A GLOBAL DESIGN EXPERIENCE: PACE

Our final foray into global programs was not a study abroad program, but it very much fits as part of global awareness. It is known as "PACE," which stands for Partners for Advancement of Collaborative Education, a consortium of companies (led by General Motors) and universities, which work together to provide students with international collaborative CAD/CAM experiences. BYU became involved in PACE about ten years ago when the consortium was mostly just about providing experiences for students in CAD/CAM. One faculty member, Greg Jensen from Mechanical Engineering, was instrumental in arranging for BYU to be invited to be a member. Gradually, as more universities joined, the consortium morphed into more of a collaborative design experience, with common CAD/CAM software acting as the thread tying all the teams together. Under Professor Jensen's direction, BYU students coordinated the efforts of twenty student teams across the globe, including teams in China, India, Korea, Sweden, Germany, Brazil, Mexico, and Australia, in designing and building a Formula 1 style racecar.[7] Subsystems were designed and built by student teams at their own universities; these systems were assembled at BYU into a working vehicle. In some instances, BYU students traveled to other universities and trained students in the use of software tools.

This project illustrated how the unique capabilities of the student body might be applied and how BYU could be a leader among engineering colleges in these types of activities. BYU was chosen as the lead university because (1) the faculty advisor was very able and committed to the project, raising $200k for the project on his own, and (2) the student team spoke seven languages and could communicate with all the other teams around the world.

[6] A. Frankman et al., "Training Internationally Responsible Engineers," 2007.
[7] Webster et al., "PACE Global Vehicle Collaboration," 2007.

QUESTIONS AND REFLECTION

As we now enter our third year, I feel a need to reflect on our experience. Here are some of my questions and observations about our experiences so far.

First, and perhaps most importantly, are these programs worth the cost? Are students learning what they need to learn and should learn through these programs? It is the case that these programs are relatively expensive both in terms of money and faculty time. Some faculty expect students to have a transformative experience as a result of these trips. That is a pretty high standard to meet and not one we apply to anything else in the undergraduate program. Nevertheless, from visiting with faculty directors and from accompanying students on these trips, I believe that many students do have transformative experiences. These can be the result of gaining a new perspective on themselves as members of a global community or arise from a new view of their possible career paths as professionals working in a global setting. In this regard, students are sometimes anxious about working outside the U.S. A study abroad experience can quiet fears and help them see this is an opportunity they should consider.

Assessment of these programs continues to be a challenge. As previously mentioned, we have a set of general outcomes for each type of format, and we ask each director to identify a set of learning outcomes specific to the particular program and format, share these with the students beforehand, use them to drive discussion and learning, and have the students self-assess if these have been reached. This kind of indirect assessment is not as meaningful as more direct measures. We have investigated the Intercultural Development Inventory[8] as one possibility here, but we are still considering whether this is a good instrument for us.

We have focused on several specific study abroad formats such as mentored travel, service learning, and international design, partly because these programs scale relatively well. When I refer to scaling, I mean that the programs can each accommodate up to about twenty students. This may not seem like a lot, but my examination of study abroad programs shows that few formats can run higher than this number, and many run at less. Choosing these formats has allowed us to ramp up relatively quickly and go from perhaps 2-3% of the students in the college being involved in some sort of global experience to around 20% of the students being involved.

The most important key to success is developing a cadre of faculty directors. Anyone who has been a director of one of these programs knows it is a lot of work, both in terms of the effort required before leaving and in terms of the hours involved while abroad. One of our greatest concerns is providing rewards and support for faculty so they continue to want to do this. In the future we may be constrained in our ability to expand study abroad opportunities by the number of faculty willing to be program directors. We have provided faculty who have started new programs with fellowships and other support. For longer programs, we encourage faculty to take spouse and family, if they desire. In the end, faculty members are driven by the reward system. Although directing a study abroad program cannot compensate for a fatal weakness in some other area of performance, it is viewed very positively in the college and has been a positive factor in rank advancements. As an

[8] Hammer et al., "Measuring Intercultural Sensitivity," 2003.

example, recently the participation of a faculty member as a director of study abroad was the "tipping factor" which resulted in a recommendation that he be promoted (the file was solid in teaching and research but otherwise would have been weak in citizenship).

As illustrated by the comments of Jim Nelson, the director of the Mexico program, the development of study abroad programs discussed here would not have happened without the support of the college leadership. College support has communicated to the faculty that we value these activities. We are in a position to direct resources to these activities and put "our money where our mouth is." We have directed about $75k per year to student scholarships and perhaps another $50k per year to faculty support.

It is reasonable to ask, "Would these programs continue if the college leadership changed?" Of course, it is not possible to answer that question definitively, but I believe they would. I have felt a shift in the culture of the college towards the attitude that these programs form an important part of the spectrum of learning activities that should exist in a modern engineering college. It has helped that the load of running these programs has been pretty evenly split among departments. Our Engineers Without Borders directors come out of the Chemical Engineering Department; China comes from Electrical Engineering; Mexico and China Mega-structures (a new program in 2008) are from Civil Engineering; and Europe is from Mechanical Engineering. This dispersion of directors has not only allowed for load sharing across the college but also provided a leavening influence as these directors are able to share their experiences with their own departments.

We have learned the importance of good student preparation before study abroad begins. Based on our 2007 experience, we are moving to have a more consistent and better structured preparation seminar for all programs. Ideally students should receive some exposure to four areas: (1) cultural issues, such as cultural diversity, ethnocentrism, communication across cultures; (2) country issues, such as a brief introduction to a country or region's history, economy, and politics; (3) study abroad issues, such as handling money, safety, and health; and 4) globalization issues, such as trade policy, outsourcing, intellectual property, and technology.

These programs are synergistic with another college goal of increasing women enrolled in engineering and technology. About 25% of the participants have been women (which is, sad to say, double the percentage of women in the college).

OTHER CHALLENGES

There have been the obvious challenges of organization and logistics associated with these programs. We could not have grown as rapidly as we have without the help of the Kennedy Center for International Studies on campus, which coordinates all study abroad travel for the university.

Challenges associated with logistics are the easy ones. We have also faced the challenge of maintaining balance in the college. As an illustration of this, I relate the following experience. Every school year begins with a week of meetings called collectively "University Conference." One day of these meetings is allocated to the college. This is one of the few times all faculty of the college gather together. As dean, I am expected to provide an overview of where we have been the past year and where we are going. Since the beginning, we have presented a consistent vision, as explained here, and tried to reinforce the vision by presenting evidence as to how the world is changing. This year, for example, I asked the faculty to read a chapter from the document, *Engineering for a Changing World*, authored by James Duderstadt, former president of the University of Michigan and former dean of engineering at the University of Michigan.

Duderstadt indicates that the goal of American engineering schools should be "to focus more on quality, producing engineers capable of adding exceptional value through innovation, entrepreneurial skills, and *global competence*" (italics added).[9] Nevertheless, this year I received feedback from one of the chairs that some of his faculty felt we were putting too much emphasis on globalization (and the other elements of the strategic plan) and not enough on research, which they considered to be a more fundamental responsibility. I laid awake all night thinking about this. Sometimes it is the case that if you "play one or two keys long enough on the piano, people may come to feel the other keys are not important." Frankly, I had thought about the potential for this situation to develop and hoped we could avoid it. In my mind there was still a strong emphasis on research; we just were not talking about it as much as we tried to move in some new directions. However, I felt I should listen to the feedback I was receiving. After much contemplation, I shifted this year's presentation to have a primary focus on research. We will need to maintain a balance in all we do in the college. Ultimately, however, I believe we will need to expand our set of learning activities.

Along these lines, I would like to relate a recent experience which occurred at the annual ASEE Engineering Deans Institute. The occasion was a session that involved an industry panel with representatives from well known high tech companies. These people were discussing what skill set they wanted in their new hires. They wanted what a number of studies have proposed: not only a strong technical understanding, but also leadership skills, communication skills, high standards of ethics, etc. During the question and answer session, one dean asked the question, "Since you are asking for students to learn more, and the curriculum is already chock full, what do you propose we leave out?" The assumption was that we could not add in more stuff without pushing something else out. The reply of the panelist was instructive: "Nothing." He wanted it all. Whereupon another dean suggested that perhaps we should look at the M.S. as the entering professional degree. This would allow for more time to cover more ground. The panelists thought this might be a possibility.

I could not help but wonder about a third possibility: Is it possible to improve our "educational productivity?" Each year, the industrial sector becomes more productive by 2-4%. This means it produces 2-4% more per unit input of labor. Is there such a thing as educational productivity? Can we produce more in four years than we have before because we are working smarter and not just harder? And, as a college, can we grow our research and improve our teaching while developing technical study abroad programs?

I have to admit I am not yet sure of the answer to this question. The "product" we produce at a university—educated students—comes with its own agency, motivation, and preparation. Surely we have not yet reached the maximum efficiency in how we approach learning and teaching. I hope

---

[9]Duderstadt, "Engineering for a Changing World," 2008.

further gains might yet be made. Yet as I look back, I am not sure if I should be optimistic. Changes in pedagogy have been slow in coming. Many of our courses are taught pretty much the same way they were twenty or even forty years ago.

UNEXPECTED INSIGHTS

I have been a little bit surprised regarding how I have come to feel about humanitarian projects. Previously I viewed study abroad as a way to learn what developed countries were doing and how to prepare our students to remain competitive in the global marketplace. However, I have come to understand that another motivation for promoting international technical experience relates to the range and scale of technological needs of humankind in the 21st century. Some of the most challenging technical problems facing the world include providing clean water, sanitation, infrastructure, energy, health care, and transportation to the majority of the world's population which lives in developing countries.

I had not anticipated how receptive donors would be to humanitarian study abroad programs. Donors like having their gifts do double duty: helping students learn and helping people have a better life. As dean I have become very active in fund raising, and previously I had not connected these efforts with our efforts in global awareness. I am convinced that if handled properly, humanitarian projects can be excellent learning experiences for students. There is often synergy with other learning outcomes such as multi-disciplinary design under a set of unusual constraints (i.e., third world constraints), teamwork, and open-ended problem solving. Students also develop an understanding of themselves as members of the world community and are often humbled by what they enjoy as citizens of the United States.

I have been somewhat surprised at the emergence and potential of discipline specific study abroad programs. Originally we were thinking (in the college office) about how to run more general programs, open to anyone in the college. This seemed democratic. But some of our strongest programs have been ones that have a specific disciplinary focus, such as the program in Mexico. In 2008 we ran two new discipline specific programs: a program in Europe focused on product design, and a program in China on mega-structures (large buildings, dams, bridges). These programs require students to take certain upper division courses and thus are only open to a small subset of students. This preparation provides a strong academic background and context for the study abroad experience. I was not anticipating the development of some of these programs when we started. They have provided good learning experiences for students and have been relatively painless to run at the college level since they are sponsored by departments.

CONCLUDING REMARKS

My hope as dean is that I might leave the college stronger than I found it. I hope our students might be better prepared to have successful careers and a satisfying life because of opportunities made available by the college. No one can see the future. But I feel quite certain that the globalization of engineering we have witnessed over the past twenty years is here to stay. And, certainly, if we are going to tackle the problems facing humankind in the 21st century, we will need engineers who can work across ethnic, cultural, and national boundaries.

ACKNOWLEDGEMENTS

Although this account has focused on my own journey in developing global awareness, I wish to acknowledge the help and support of my associate deans, John Harb and Spencer Magleby. We have been united and worked together in this adventure. I also appreciate the willingness of some faculty to act as program directors. We have been fortunate that a few have had a passion for doing this. I also express appreciation to the dedicated personnel at the Kennedy Center for International Studies and in particular to Chelita Pate, who has helped arrange our trips.

REFERENCES

Christensen, C. M. *The Innovator's Dilemma: When New Technologies Cause Great Firms to Fail,* Harvard Business School Press, 1997. 221

Duderstadt, J. "Engineering for a Changing World." The Millennium Project, University of Michigan, 2008. `http://milproj.dc.umich.edu/publications/EngFlex_report/index.html`. Accessed July 2010. DOI: 10.1007/978-1-4419-1393-7_3 231

Engineers Without Borders. `http://www.ewb-usa.org/` Accessed July 2010.

Frankman A., J. Jones, W. V. Wilding, and R. Lewis. "Training Internationally Responsible Engineers." Paper presented at the 114th ASEE Annual Conference and Exhibition, Honolulu, HI, United States, 2007. 228

Friedman, T. *The World is Flat: A Brief History of the 21st Century.* New York: Farrar Strauss and Giroux, 2005. 221

Hammer, M.R., M.J. Bennett, and R. Wiseman. "Measuring Intercultural Sensitivity: The Intercultural Development Inventory." *International Journal of Intercultural Relations* 27 (2003): 421–443. `http://www.intercultural.org`. Accessed July 2010. DOI: 10.1016/S0147-1767(03)00032-4 229

National Academy of Engineering. *The Engineer of 2020.* Washington DC: National Academy of Engineering Press, 2004. 223

Parkinson, A. "Engineering Study Abroad Programs: Formats, Challenges, Best Practices." *Online Journal for Global Engineering Education* 2, no. 2 (2007). `http://digitalcommons.uri.edu/ojgee`. Accessed July 2010. 224

Webster, M., D. Korth, O. Carlson, and C. G. Jensen. "PACE Global Vehicle Collaboration."Paper presented at the 114th ASEE Annual Conference and Exhibition, Honolulu, HI, United States, 2007. 228

CHAPTER 11

The Right Thing to Do: Graduate Education and Research in a Global and Human Context

James R. Mihelcic

BACKGROUND

Back when I was an undeclared first-year engineering student, I remember getting a call during the spring semester from my father. This was just after first-year students were required to select our engineering major. My father, a man of business who saw engineering as a first step towards a second degree in business or law, asked on the phone, "So what did you select, mechanical, chemical engineering?" I remember sheepishly answering, "No, I selected environmental engineering." After some silence on the phone, he responded back, "What do you want to do the rest of your life, paddle a canoe?"

Several years later, my father became a huge supporter of my selection of environmental engineering as a career, but at that particular moment, I did not receive the support a young person yearns to receive as they enter uncharted waters. But even at this critical moment in my career, I see how I was merging both personal and professional interests, in this case, engineering and protecting the environment.

Looking back on that moment and the three decades that followed, I can say that standing up to a loving parent is not an easy thing to do. Standing up to colleagues and administrators who question or block your ideas to innovate engineering education is also not easy. However, I can say that combining a passion for the environment and issues of social justice, and an expectation of service with engineering, was always the right thing to do. This ultimately led me to transform some of my teaching portfolio and all of my research into sustainability, now embraced by the majority, and issues of sustainable development and humanitarian engineering through the Master's International program.

In 1997, I did what I term here "the right thing to do" by creating a partnership that allowed graduate engineering students to integrate education and research with training and service in the Peace Corps. What started from modest beginnings has grown into a roaring success of educational reform, external funding, peer-reviewed journal articles, and textbooks, fueled by a diverse group of motivated students who crave life-long learning. In those early years, students took a leap of faith that allowed them to integrate their engineering education with language and cultural immersion and research in a developing world setting. Most of us know that doing the right thing is personally and professionally fulfilling. What we do not all know is that it can also be integrated into university measurements of performance.

After over ten years of building this program, I switched universities, traveling south on US41 to Tampa and the University of South Florida. We still have engineering students devoting two-plus years to service and research in the developing country, but more importantly, the program has matured and embraced institutional strengths specific to my new location. Now students study how issues of culture, geography, language, gender, poverty, health, and technology influence the solutions to what are becoming increasingly complex global problems.

To date, the thirty-credit Master's International (MI) Program[1] has allowed graduate engineering students to integrate their education and research with eleven weeks of training and two years of engineering service in the Peace Corps. The program not only provides an in-depth understanding of their discipline but importantly, the relationship of engineering to the environment, health, social needs, language, geography, and culture. In the process, students also develop solutions with the added constraints of the world's economic, social, and environmental limitations. The program is also integrated with a sixteen-credit *Graduate Certificate in Water, Health, and Sustainability* that requires curricular breadth and focused coursework in global public health, applied anthropology, and sustainable appropriate technology.

Initially, students took a leap of faith with me; in fact, many times they jumped first into the abyss despite warnings of uncharted territory and an improper path to doctoral studies or employment in traditional engineering sectors. Students have commented that it was my vision and enthusiasm that kept them going through the hard times. I would counter that it was equally their vision and enthusiasm that kept me going through similar times. They were truly the students who craved the concept of "life-long learning" that many of us only hear about before an impending ABET accreditation visit.

MASTER'S INTERNATIONAL – PROGRAM BASICS

In those early years, I had a stated goal of the Peace Corps Master's International program to educate engineers who are not only grounded in the fundamentals of engineering, but they are also educated and trained in the many critical, non-technical skills that are required of today's engineer. This partially resulted from my involvement in ABET assessment where I constantly heard employers request more non-technical skill training of our graduates and our faculty resist making substantial

[1]Master's International Program, http://cee.eng.usf.edu/peacecorps

changes. Over time I developed a vision with my colleague Linda Phillips of "a world where all have access to sanitation and sufficient safe water, where all children are able to learn in well-built classrooms, where families no longer suffer from disease, starvation and poverty, where renewable energy has replaced fossil fuels, and importantly, where colleges and universities are part of the solution."

A stated goal of the program is to educate engineers in the value of community service. The program is also integrated with educational initiatives in sustainability[2] and the three goals of the Peace Corps: (1) to help the people of interested countries in meeting their need for trained men and women, (2) to help promote a better understanding of Americans on the part of the peoples served, and (3) to help promote a better understanding of other peoples on the part of Americans. Two of these three goals are directly related to cultural integration, an important component of most study abroad experiences and something I value as an educator and mentor.

It takes a student approximately three and a half years to complete the program coursework, service, and research requirements. Students first spend two to three semesters on campus taking graduate coursework before departing for their two-plus years of overseas training, service and research (Figure 11.1). Peace Corps service can encompass all aspects of an engineering project.[3] I observed in the students over time that they tended to not be in a hurry. What I mean by this is that they saw there was plenty of time to work after graduation, and they saw a benefit in combining two passions of theirs: graduate school and Peace Corps service.

THE RIGHT THING TO DO

Universities will always value excellence in service, education, research, and scholarship. However, many times they often ignore *doing the right thing* and taking on new initiatives unless they are associated with a large donation or government-sponsored external funding. In the early 1990s, I recognized that the societal aspect of engineering was not as strong of a focal point in my teaching, research, and scholarship. I was also interested in greater integration of my personal pursuits that involved issues of the environment, justice, caring, and the importance of engineering involvement in decisions being made in our local communities.

At this time, my research colleagues had just received a large competitive grant from EPA to establish a national research center. The center's goal was to develop clean technologies that would support pollution prevention activities. The center's interdisciplinary focus brought together the disciplines of environmental, chemical, and mechanical engineering. During the same period, I was also involved in a real David versus Goliath story, where a grassroots community group was attempting to halt a proposal to construct a large pulp and paper manufacturing plant that was

[2]Mihelcic and Hokanson, "Educational Solutions: For a More Sustainable Future," 2005; Fuchs and Mihelcic, "Engineering Education for International Sustainability," 2006; Hokanson et al., "Educating Engineers in the Sustainable Futures Model with a Global Perspective," 2007; Trotz et al., "Non-traditional University Research Partners that Facilitate Service Learning and Graduate Research for Sustainable Development," 2009.
[3]Mihelcic, "Educating the Future's Water Professional," 2004; Mihelcic et al. "Integrating a Global Perspective into Engineering Education and Research," 2006.

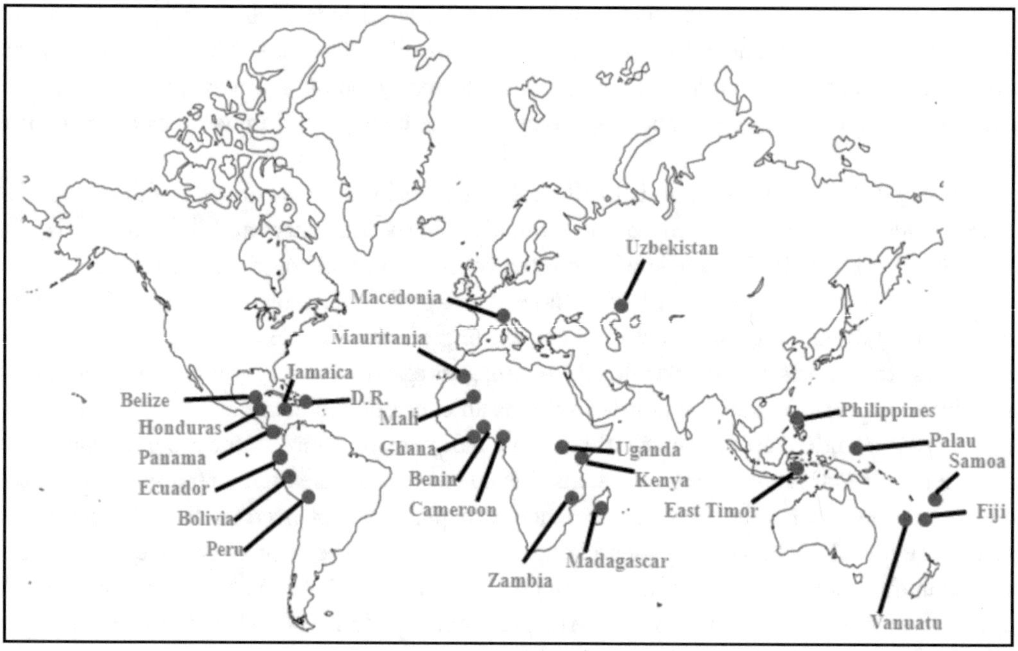

Figure 11.1: Locations where my students have combined their education and research .

sized well over the area's ecological carrying capacity (interestingly, in our story David actually won). In this case, a small community group came to me for technical advice to counter the enormous resources the paper company and pro-industry state regulatory agency had available.

At this moment, engineering and academics were still enjoyable. I was, however, aware of the increasing complexity of environmental issues and challenge of educating students to be prepared to adequately address them. Importantly, I also realized that those general education classes I had enjoyed as an undergraduate were, indeed, very important to the student in the workplace and the citizen in the community. They thus became something more to me than courses a few colleagues might complain about when curricula reform or ABET accreditation came up at a department meeting.

I began to meet with faculty in social sciences and read literature from fields outside of engineering. I cannot remember the specific types of material I was reading at this time. Some were academic, suggested by my colleagues in social sciences, and some were located in the mainstream, international, or progressive press. All of them tended to have a common theme located at the intersection of people and the natural environment. This all took some time because I still had my conventional engineering literature to read.

THE BEGINNING

In 1996, I was having a coffee with a colleague while glancing at a university announcement that the Peace Corps would be on campus for a signing ceremony to start something called a "Master's International Program" housed in Forestry. I turned to my colleague and said, "You know, we should do that." At that time, I had no idea what that meant as an educator, researcher, or scholar. My colleague, considered wise by most university standards, counseled me that even being newly tenured, being an associate professor meant there were other and more important things to focus my efforts on— things like more doctoral students and research papers in scholarly journals. Looking back on this moment, I would say it was probably good I was more focused on some unknown possibilities than on the enormity of what I would encounter.

I tend to pursue things that integrate my heart and personal convictions with the structure of my engineering brain. I was always willing to devote the same time to an undergraduate in need as to a doctoral student of great promise. And I was always willing to assist a small community group that needed some technical advice as much as a larger and better funded entity that requested my assistance. So I quickly set up a meeting with the Peace Corps representatives who immediately expressed excitement in creating what would become the first Peace Corps Master's International Program in Engineering.

All I had to do was figure out how to incorporate the one semester of overseas Peace Corps training, a minimum of six semesters of Peace Corps service, and overseas research into the existing graduate program structure. Then I would just have to recruit, train, and mentor students.

A few years later a colleague from a prestigious university commented to me after we had discussed the program, "You mean the Peace Corps provides you no research dollars which end up reported as university research expenditures?" I nodded my head in agreement to which this colleague responded, "You mean you did this because *it was the right thing to do?*" Again, I nodded by head in agreement.

In terms of timing, the birth of the program was before the creation of organizations such as Engineers without Borders-USA or Engineers for a Sustainable World. I have often told students that when they are older, they will look back on that leap of faith and no longer see themselves as part of a minority. Instead, a model that once was uncommon will now have become the norm in engineering education, perhaps even a university movement. This vision of "what will be" is a powerful motivator for me, faculty colleagues, and students.

THE EARLY YEARS

In regard to the Master's International program, the first year consisted of setting up the program and recruiting, so there were no students enrolled. At the beginning, I required students take four credits of preparatory work that would prepare them to function as an engineer in the Peace Corps setting (now, at the University of South Florida, this has expanded to nine credits). I could rely on my colleague in forestry, who was a returned Peace Corps volunteer and directed a similar program, to teach a two-credit course focused on community participation and planning. And as I began to develop a two-credit course on appropriate technology, I located dozens of books and technical briefs that were written by others immersed in overseas development. I would also listen to any person who had done development work. The challenge was to synthesize all this information into something practical that was rigorous and worthy of graduate credit.

The second year brought the first student with a degree in earth systems engineering, and the third year brought the second student with a mechanical engineering degree. At that moment, there was not much to brag about in an end-of-year review, but my regular teaching and research duties were going fine, and the two students in the program were excellent and willing to share their stories and knowledge with me, even if they were in Africa and Central America at the time. Other small indirect positive things happened, like for example, a few domestic doctoral students entered our graduate program after first learning about us through the Peace Corps.

In all honesty, I have to say at times things were difficult for me, balancing my conventional teaching and research programs with this new initiative that had a rather steep learning curve and little support from administrators and most of my colleagues. In fairness to my colleagues, they were buried in the normal duties of being a professor. Looking back, I think what kept my trajectory moving forward was the students and my sense that conviction and values were merging with my profession. In our hearts, the students and I knew we were doing the right thing. And I knew that I never had a complaint about the students or the knowledge they were creating.

At first, it was hard for me to let go of my conventional graduate class. I wanted to teach it; after all, it always had a healthy enrollment and was the bank vault of my collective research knowledge up to that point. But as all academics know, time is finite, and I only had time to teach a limited number of courses. Also, the conventional graduate class focused on a reactive engineering approach of remediation, and I wanted to focus on proactive approaches related to prevention of pollution. In addition, I had a new class that had to be taught and too many resources to sift through and teach from. Fortunately, after fourteen years, we now have the book I wanted in those early days. It is based on collective knowledge of the first fifty students we had in the program.[4]

The students, very loyal to the program's vision, persevered through their overseas assignments, conducted their required research through natural disasters and personal health issues, and continued to transfer knowledge back to me and other students that I could use to update my course notes and advising skills. Over this period, students showed a significant amount of independence as they all identified several research topics during their first six months in the field. All they needed was me to assist them in refining the topic and place it in a context that would be acceptable to a graduate committee and outside peer reviewers as I challenged all of them to disseminate the knowledge they were generating.

As a critical mass of students developed, by year three and four, interest had increased to new class of six or so students per year. Then the program exploded to annual enrollments of ten-twelve students along with some other undergraduate and graduate initiatives related to sustainability that I was developing. In all honesty, the timing of student interest was related I think to an awakening

---

[4]Mihelcic et al., *Field Guide in Environmental Engineering for Development Workers,* 2009.

that was occurring amongst the Millennium generation, who were now expressing stronger desire for programs related to environmental degradation, climate change, poverty, justice, renewable energy, and serving others.

Now the program was full of students; some were new and eager to learn, others were seasoned and returning from their overseas assignment to defend their research and tell me and the new students what life as an engineer in a developing world community consisted of. I would spend a significant amount of time with returning students. We talked about many things: their overseas projects and living conditions, how the program had prepared them or what I could do better, and what their next professional step was going to be. And if I had some free time, I would pay out of my pocket to travel overseas and spend time with them in their communities and work on their engineering projects.

More and more of my time was now taken up responding to program inquires, or advising students who were in various stages of the program. I tried to engage other faculty to learn about the program and advise students. But except for a few, most just stood aside in their conventional education and research world. I am happy to say though that in my new academic setting, I have a chair, dean, and a large number of faculty colleagues who see supporting and advising these students as an integral part of their education and research.

ISSUES

KNOWLEDGE CREATION AND ACADEMIC ACHIEVEMENT

I always understood that the students were creating knowledge. However, it took me several years to figure out how that knowledge could be disseminated, either through traditional or non-traditional methods. Fortunately (or unfortunately), I had my conventional research program at this time to keep my publication numbers high. It is now clear that the students could be challenged to do high quality research of community and global significance that was publishable in peer-reviewed literature. In all honesty, I was truly not sure what the form of dissemination would be when I started this program. I just challenged the students and myself from day one to find a suitable outlet for dissemination.

Eventually, significant external NSF funding for research and scholarship —framed by the experiences of students— followed and supported other undergraduate and graduate students. In fact, as I wrote the first draft of this manuscript, I was returning from an extended research trip in Bolivia, supported by the National Science Foundation, with a group of U.S. doctoral and undergraduate engineering students who team with Bolivian undergraduate engineers to work on sustainability problems of water and sanitation in an international context. Even this program has now evolved to team students from engineering, anthropology, and our undergraduate Honors College.

I have learned though that it is quite common that international and developing world research and data collection require more time. It is not as simple as telling a student who has returned from the field that they need to repeat an experiment or collect a missing piece of data. Their "laboratory" could be thousands of miles away, via a long plane ride and bumpy bus ride or boat trip. Their study community may have encountered some natural disaster or key personnel may have migrated to the city or even immigrated to another country.

As an example, I have been researching pathogen destruction in composting latrines for six years in Panama with three different students. Our work includes field measurements on fundamental mechanisms of pathogen destruction as well as studies on human behavior in regard to latrine operation. Each student has provided incremental knowledge, so now the four of us have finally obtained sufficient data to publish a peer-reviewed journal publication with two colleagues in global public health.[5] This time frame may not be appropriate for a junior faculty member who needs to demonstrate in six years the involvement of several graduate students with multiple papers.

I know from personal experience that it may also seem daunting to learn a new body of literature and move a research publication through peer review when perhaps the reviewers are skeptical of the quality of research coming from a village in Africa. Promotion committees and department chairs may also not fully understand a new body of literature that is more interdisciplinary and perhaps has a more applied audience of nongovernmental organizations. Fortunately, motivated students, if pointed in the right direction or provided a few atypical key words, will seek out new information and journals.

HOW MY THINKING AS AN ENGINEER EVOLVED

In the process of the program's evolution, I began to see the similarities and complexities of the scale of problems, as they evolved from the laboratory to a community, then to a watershed, and, finally, a continent and the planet. This required that I redefine what all this meant to me and my students as an educator, researcher, scholar, and inhabitant of a planet that had a human population heading towards nine billion. Obviously, this change did not come with complete ease. As professors, we know that time is the one thing that is finite in our lives. However, as I look back on these years, the personal and professional time committed to this endeavor seems great from a cumulative standpoint, yet the joy created from my initial investment has paid much greater dividends.

Over time, I observed how my approach to solving problems changed because, just like the students, I also was placed in unfamiliar territory. For example, I quickly learned that the problems that students encountered (and the resulting solutions and designs) were different than how I had been trained to solve problems. Being in unfamiliar territory initially adds complexity to problem definition and problem solving, but it also open up degrees of freedom if embraced early in the design phase while also providing a rewarding human dimension.

From the students, communities we partnered with, and the problems we focused on, I learned many things that I was able to incorporate into updated course notes for all the classes I was teaching, whether I was preparing a student to work in the U.S. or Latin America. This understanding that I needed to add to the students' professional portfolio of health sciences, anthropology, and geography was one motivation for me to join a new university which had strong research and education programs in these areas as well as a track record and commitment to being globally engaged.

[5]Mehl et al., "Pathogen Destruction and Solids Decomposition in Composting Latrines," 2010.

As just a few examples, the word *project* now became more than a physical structure and correctly included the social setting where the project is located and the people who operate, manage, and benefit from the project. *Beneficiaries* became real people with real concerns whose input was critical to a project's success, not simply, a nuisance at a public meeting. The importance of *gender* in developing an appropriate solution became apparent, and I clearly saw how *health* was integral to sustainable development and was missing in most engineering design. I also observed how the social and cultural aspects of a problem could be combined with a life cycle thinking approach espousing that a technological solution is best assessed with what we termed the *five pillars of sustainability*: (1) socio-cultural respect, (2) community participation, (3) political cohesion, (4) economic sustainability, and (5) environmental sustainability.[6]

As I became more knowledgeable about what skills a successful student needed and further developed my thoughts on broader issues of sustainable development, what was initially a four-credit core coursework requirement in those early years developed into nine credits at the University of South Florida. These nine credits embraced three courses: global public health assessment, research methods in applied anthropology, and appropriate sustainable development engineering. I also became aware of how working in the developing world communities reopened my eyes to the vast amount of indigenous knowledge that could be transferred to solve problems at home and abroad.

LEARNING LANGUAGE

I remember as an undergraduate how I wanted to take a foreign language and "ABET" would not allow me within the confines of my engineering degree. I also remember how I wanted to do a "study abroad" in St. Petersburg and there was no room in the curriculum for the courses or funds to assist the semester overseas. I now am an aggressive supporter of allowing students (even my doctoral students) time and flexibility to become proficient in a foreign language. Though my foreign language skills remain a source of frustration and can contribute to my sense of isolation when I travel abroad, in contrast, my sense of pride is overwhelming when I experience my MS and Ph.D. students communicating in another language in both social and technical settings. To date, over 65% of our Master's International students have become fluent in French or Spanish, and the remaining students learn another language (e.g., Patois in Jamaica, Bislama in Vanuatu). Many of the students actually become proficient in two languages during their Peace Corps service (e.g., French and Bambora in Mali, Spanish and Nobe in Panama).

DISCIPLINE DIVERSITY

One of the most exciting things about a risky and different type of program is the diversity it brought to our classrooms and my research group. This diversity became apparent in the form of gender, race, geographical home, and previous degree. Important to the engineering discipline and discussion

---

[6]McConville and Mihelcic, "Adapting Life Cycle Thinking Tools to Evaluate Project Sustainability in International Water and Sanitation Development Work," 2007.

in the classroom or hallway, I have been amazed at the wide range of our students' first degrees (Figure 11.2).

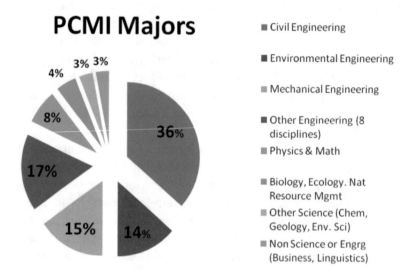

PCMI Majors

- Civil Engineering
- Environmental Engineering
- Mechanical Engineering
- Other Engineering (8 disciplines)
- Physics & Math
- Biology, Ecology. Nat Resource Mgmt
- Other Science (Chem, Geology, Env. Sci)
- Non Science or Engrg (Business, Linguistics)

3% 3%
4%
8%
36%
17%
15% 14%

Figure 11.2: Majors of Peace Corps Master's International students prior to entering the graduate engineering program.

THE LITTLE THINGS

Was I prepared during my education for the ups and downs that make up daily life in the communities the students work in and I visited? Was I prepared as an educator to discuss the ethical dilemma of what was more valuable, using technology that improves health but increases human populations or protecting ecological systems under siege from human population growth? The answer is, of course, no, but it made for classroom discussion.

It can also be easy to forget a student who is thousands of miles away and only has computer access a few days a month. That student may also be in ill health or have fluctuations in their mental state, which can include feeling isolated in one instance, or yearning to be alone at another. Therefore, I changed my daily routine to be responsive to students' requests. This may mean responding to email requests within twenty-four hours, involving one's on-campus conventional research group to help track down literature, or quickly packaging some research supplies for the one-month journey to a student's site. With the students' long stays overseas, mentoring might now consist of writing a personal letter or mailing a care package. These were all not easy things to fit into the hectic life of a university professor (or to explain to a department chair).

When students return from the field to complete and defend their thesis, it is critical they transfer knowledge to other students and faculty advisors, not only on field research methodologies that work, but also importantly, issues of daily community life and their work as engineers. This method of knowledge transfer is applicable to language and cultural integration, professional practice, obstacles encountered and methods to overcome them, along with issues related to research methodology.

Inviting students into my home has been one of my greatest pleasures. These students are full of once-in-a-lifetime experiences and eagerly share them. Their personal and professional growth over the two-year period is outstanding. Being surrounded at a dinner table with a group of young adventurists telling stories over an exotic meal and a bottle of wine is the ultimate reward. And then visiting them and experiencing these stories firsthand; peering into an active volcano and hearing the thud of molten rocks hitting the ground on the opposite ridge as they return to earth after being hurtled skyward, taking a dugout canoe back to sleeping quarters at night with alligator eyes shining in the glare of a flashlight and hoping to catch a glimpse of a jaguar, sharing a meal with a village chief and his family, celebrating with a community when the water is turned on, and being invited into a home where a family with nothing gives so much.

One of the most satisfying and exhilarating little things is the correspondence with students, via a letter scribbled in small print inserted in an envelope covered with the art of overseas stamps, or more likely today, a blog posted with exotic photographs. I always tell students that they have better stories than anything cable television could ever provide. Yet while many of these stories are of the triumph of the human spirit or once in a lifetime experiences in some far off village, some have a theme of tragedy that is always present in the human race, especially in communities where poor health and malnutrition may be common occurrences.

ALTERED RELATIONSHIPS WITH STUDENTS

I have traveled overseas with undergraduate and graduate students. A small group might be one or two graduate students. On the other extreme, this past summer in an isolated part of Bolivia, our research group was thirteen students, of which twelve were female and students were equally split between Americans and Bolivians.

It was not until after this trip that I suddenly realized how the *space* and *formalities* between the faculty member and student may be greatly reduced compared to the space and formalities that exist in the hallways outside the office and classroom. This occurs for several reasons, some being closeness of living quarters, sanitation, clothes washing areas, and even the cramped space in a taxi bouncing down a dirt road. For some, this can turn into an uncomfortable situation. Space and formalities are also reduced because of the reality that there are occasions when you and the students not only feel ill, but also visibly show your illness. Disease, change in diet, change in climate, motion sickness, and even culture shock can all cause illness. Furthermore, while in Bolivia it is culturally acceptable to greet your female student with a handshake and kiss on the cheek, returning to the U.S., this type of exchange between a faculty and student would be inappropriate.

This situation is not negative though; it opens up doors for education and mentoring that are transformational for both faculty and students. Students may now see the job of a faculty member as more human and satisfying than the rigid, high level of work they sometimes observe when we are sheltered in our offices and classrooms.

BEYOND REGULAR WORK

Culture shock is an inevitable occupational hazard of living and working overseas. The U.S. State Department describes it as the three stages of psychological disorientation people experience when they move for an extended period into a culture markedly different from their own. During culture shock, a positive mindset, which is initially embraced upon arrival, can soon be replaced with frustration, distress, and depression because of what appear to be irritating differences between cultures that impact your day-to-day life and work. Issues of language can make you feel isolated, in both professional and social situations. Faculty need to understand they will go through these stages; it is part of the reality of working for longer periods overseas.

All these problems are further complicated by difficulty staying in touch with *regular work*. Phone and Internet use may become very limited, not only due to access, but also, to issues of phones and power not working on a regular basis. This can be very frustrating as one perceives email and regular mail piling up. Email demands an immediate response; the reality of extended travel overseas may be that responses are rarely immediate.

Other bureaucratic expectations are problematic as well. I used to even dread returning back to a research accounting office that had already scolded me, on several occasions, for the lack of formal receipts you are provided on a trip. I have thought of asking (in my poor Spanish) the old man moving a boat manually with a pole across the wide and fast-moving Rio Beni for a receipt for the fifteen-cent trip, just to provide me with some humor to the whole situation.

One time in the small village of Sapecho in Bolivia, I was with students working with the local water committee on how to better operate their wastewater treatment system. We proposed the idea that purchasing a pair of ducks might reduce the need to manually clean unwanted plant growth on the top of a wastewater treatment lagoon. When the *campesinos* embraced the idea and said they could obtain the ducks, I pulled out eighty Bolivianos from my wallet and they departed excitedly up the trail through a banana plantation to purchase the ducks. Standing there in the hot sun, instead of celebrating this innovative solution made by my engineering undergraduates who had researched several appropriate solutions on their own, all I could immediately think about was how I was going to explain this purchase to research accounting. Even in a time of educational triumph, the reality of university bureaucracy can dampen one's enthusiasm.

I have learned that all these barriers and negatives will arise, and they may even bring you down for a short term. However, they need to be tempered with the wonderful knowledge you gain, expressions of wonder and innovation on the students' faces, and the professional arcs students traverse after graduation. After graduation, a large percentage have continued their education with doctoral studies, others remained overseas to work with nongovernmental organizations (NGOs), and some migrated to Capitol Hill to work on policy issues that impact science and technology. One is even in law school. Others work as engineers in the "real world," but many on projects that now have social meaning, such as restoring the Florida Everglades, providing water and sanitation services on Native American Reservations, designing green buildings, and implementing natural low impact hydrological controls for stormwater management. The program thus opened my own eyes to the many opportunities available to those with an engineering degree, a secret that many of our university career centers hide from us and our students.

ADDRESSING THE PROBLEM OF SCALE

LIMITS TO GROWTH

Visions often clash with the on-campus reality. University administrators may respond to requests for space and other resources that are supported by large grants of countable dollars. A dean may get excited and respond to issues of leadership and innovation, but he/she would be more comfortable if they were framed in terms of traditional manufacturing partners that provide external funding to the college. The frustrating reality is that they may respond less quickly to longer term issues of curricular reform, service, and creating more loyal and connected alumni. Many engineering administrators are also risk averse, which can limit their excitement for a different educational or research vision. And the input of faculty members who view change as positive (or inevitable) can be continually overruled by those who develop barriers to resist change. These faculty colleagues, chairs, and deans you may encounter—educated and immersed in a more traditional way of educating engineers and unwilling to admit, or unaware, there is a rising tide of student energy for change, and, more importantly, a rapidly changing world.

The many aspects of the *limits to growth* were perhaps the greatest difficulties I encountered at my former institution and were a regular source of frustration. Since I created the Master's International program, I have moved on to what is clearly a more nurturing environment. Importantly, international efforts directly fit Goal 3 of the university's Strategic Plan (which is followed by the administration): *Expanding local and global engagement initiatives to strengthen and sustain healthy communities and to improve the quality of life.* The program also takes advantage of university strengths that include an international focus, sustainability, water science and engineering, renewable energy, global public health, and applied anthropology.

From my new chair, to the dean of engineering, the various vice presidents, to the Provost and President, everyone is onboard with achieving this strategic goal. In fact, during one of my early conversations with the Dean of International Affairs, she asked if I could work with her to have 100% of engineering students have an overseas experience. As I acknowledged her with a loud yes, I was ecstatic I was someplace where an administrator had set the bar so high. This did not worry me; in fact, it invigorated me given the reality that only a few percent of engineering students currently have any study abroad experience.

Just as there are limits to growth placed on our human population and patterns of consumption, there are probably limits to growth we need to understand in our profession. There is clearly a lesson learned here: programs can sprout and grow, but to flourish and change, the culture across a department and campus requires administrative support at all levels, not just a few faculty and students. The international process thus needs to become part of the core curriculum, so the faculty time and effort is valued by others in the department and is incorporated into workload allocation.

Proposing changes to engineering curriculum committees can take time and be one of the most frustrating aspects of engineering education. However, I believe international engineering programs align strategically with national goals that have documented the rapid change occurring in the world and associated implications for engineering education, research, practice, and importantly, the economic competitiveness of our nation.[7]

Time after time, at my previous institution, in department meetings and the hallway, traditional colleagues suggested to me that by incorporating a human dimension or sustainability into a technical problem, or internationalizing our curriculum, the rigor of the solution was somewhat diminished. To one such comment, I remember replying, "Are we here to educate the engineer of 1980 or the engineer of 2020?" In this simple response is perhaps one key point, that doing the right thing means embracing educational reform and looking forward, not backwards. It also brings up an important point, that incorporating a human dimension and caring aspect is just as important for the student working in the U.S. as it is for a student working abroad.

EXPANDING THE SCALE

One item I could never have envisioned is the impact this one program has had on internationalizing other aspects of my teaching and research. All my doctoral students have a skill set that defines them as an engineer, but they also have field experience or application that connects their research with a human dimension, or some distant part of the world. All are pursuing fluency in at least one additional language. Most research projects I submit for funding now have an international component to them.

Even a new engineering textbook that we published last year integrates several learning objectives that are of a caring and human dimension. The book also integrates issues of justice with engineering fundamentals and design and has a global perspective.[8] Interestingly, all the peer reviewers embraced and supported these concepts and their integration into an engineering textbook. And as I make the final edits on this chapter, the book is already in its third printing in a period of nine months.

Expanding the scale thus involves not only expanding the program so it allows more students to take advantage of the experience, but, importantly, also having students bring their experiences back to campus so that faculty can transfer the students' knowledge and experiences back into the classroom, into books we write, and into our research. This will allow those who cannot travel with us to experience the richness of the world and a new way to approach problem solving.

---

[7] See, for example, the National Academy of Engineering, "Educating the Engineer of 2020," 2005; American Society of Civil Engineers, "The Vision for Civil Engineering in 2025," 2007.

[8] Mihelcic and Zimmerman, *Environmental Engineering: Fundamentals, Sustainability and Design*, 2010.

CONCLUSION

Looking back, doing the right thing is not a bad thing to do. In fact, it got me through those early years. It also allowed me to overcome the culture shock of traveling abroad for extended time periods, a backlog of email, and a low-grade illness that accompanied me home. I believe it is also what makes our job as educators and mentors, as well as life, so special and enjoyable.

Many years ago, I asked my second graduate what he had learned from his experience. Insightfully, he answered that along with gaining valuable engineering skills, he also learned what it was like to put engineering into practice while taking into consideration the social, economic, and environmental limitations of the developing world. I could not have come up with a better description of our efforts. These gems and memories are the important driver for me as my colleagues and I move forward, surrounded now with a supportive Strategic Plan and a large supporting cast of students, faculty, and administrators. They are all part of a movement, a movement that has embraced *the right thing to do*.

ACKNOWLEDGEMENTS

I thank the Peace Corps and the communities we have partnered with for their support and passion for effective change. Part of material presented here is base upon work supported by the National Science Foundation under Grant Nos. (OISE 0623558 and 0966410) and the State of Florida through the 21st Century Scholars Program. I especially thank the many students and Karen who have taken the leap of faith with me. You have all enriched my life.

REFERENCES

American Society of Civil Engineers. "The Vision for Civil Engineering in 2025." Reston, VA: American Society of Civil Engineers, 2007. 248

Fuchs, Valerie J. and James R. Mihelcic. "Engineering Education for International Sustainability: Curriculum Design under the Sustainable Futures Model." Paper presented at the 5th Annual ASEE Global Colloquium on Engineering Education, Rio de Janeiro, Brazil, 2006. 237

Hokanson, David R., James R. Mihelcic, and Linda D. Phillips. "Educating Engineers in the Sustainable Futures Model with a Global Perspective: Education, Research & Diversity Initiatives." *International Journal of Engineering Education* 23, no. 2 (2007): 254–265. 237

Master's International Program in Civil & Environmental Engineering, University of South Florida `http://cee.eng.usf.edu/peacecorps`. Accessed May 21, 2010.

McConville, Jennifer R. and James R. Mihelcic. "Adapting Life Cycle Thinking Tools to Evaluate Project Sustainability in International Water and Sanitation Development Work." *Environmental Engineering Science* 24, no. 7 (2007): 937–948. DOI: 10.1089/ees.2006.0225 243

Mehl, Jessica, Josephine Kaiser, Daniel Hurtado, Daragh A. Gibson, Ricardo Izurieta, and James R. Mihelcic. "Pathogen Destruction and Solids Decomposition in Composting Latrines: Study of Fundamental Mechanisms and User Operation in Rural Panama." *Journal of Water and Health* (2010). 242

Mihelcic, James R. "Educating the Future's Water Professional." *Water Environment Technology* 16, no. 9 (2004): 86–92. 237

Mihelcic, James R. and David R. Hokanson. "Educational Solutions: For a More Sustainable Future." In *Environmental Solutions*, edited by Nelson L. Nemerow and Franklin J. Agardy, 25–58. Elsevier, 2005. 237

Mihelcic, James R., Linda D. Phillips, and David W. Watkins. "Integrating a Global Perspective into Engineering Education and Research: Engineering International Sustainable Development." *Environmental Engineering Science* 23, no. 3 (2006): 426–438. 237

Mihelcic, James R., Lauren M. Fry, Elizabeth A. Myre, Linda D. Phillips, and Brian D. Barkdoll. *Field Guide in Environmental Engineering for Development Workers: Water, Sanitation, Indoor Air.* Reston, VA: American Society of Civil Engineers (ASCE) Press, 2009. 240

Mihelcic, James R. and Julie Beth Zimmerman. *Environmental Engineering: Fundamentals, Sustainability and Design.* New York: John Wiley & Sons, Inc., 2010. 248

National Academy of Engineering. "Educating the Engineer of 2020." Washington, DC: National Academies Press, 2005. 248

Trotz, Maya A., Helen E. Muga, Linda D. Phillips, Daniel Yeh, Amy Stuart, and James R. Mihelcic, J.R., "Non-traditional University Research Partners that Facilitate Service Learning and Graduate Research for Sustainable Development."Paper presented at the 2009 World Environmental & Water Resources Congress, Kansas City, MO, United States, 2009. DOI: 10.1061/41036(342)203 237

Authors' Biographies

KACEY BEDDOES

Kacey Beddoes is a Ph.D. student in Science and Technology Studies at Virginia Tech. Her current research interests are interdisciplinary studies of gender and engineering education research and international engineering education. She serves as Managing Editor of *Engineering Studies* and Assistant Editor of the *Global Engineering* series at Morgan & Claypool.

GARY DOWNEY

Gary Downey is Alumni Distinguished Professor of Science and Technology Studies and Affiliated Professor of Engineering Education and Women's and Gender Studies. A mechanical engineer (B.S., Lehigh University) and cultural anthropologist (Ph.D., University of Chicago), he is author of *The Machine in Me* and co-editor of *Cyborgs and Citadels*. He serves as editor of *The Engineering Studies Series* (MIT Press), *Global Engineering* series (Morgan & Claypool Publishers), and *Engineering Studies* journal (Taylor & Francis/Routledge). He is co-founder of the International Network for Engineering Studies as well as founder and co-developer of the Engineering Cultures course (ranked #2 of 190 multimedia contributions to www.globalhub.org in 2010). An ethnographic listener interested in engineering studies, he researches practices of knowledge in service.

GAYLE ELLIOTT

Gayle Elliott earned B.S. and M.S. degrees from University of Cincinnati, and is currently Associate Professor in the Division of Professional Practice. She is responsible for the University's International Co-op Program, and for placing mechanical engineering students in co-op jobs in the US. Initially part of the College of Engineering, Gayle has worked with the International Engineering Co-op Program since 1993. In 1998 she created and began working with similar programs in the Colleges of Applied Sciences, Business and the College of Design, Architecture, Art and Planning. She has extensive experience developing international exchange programs and is an active member in several international engineering education organizations and projects.

LESTER A. GERHARDT

Lester A. Gerhardt received his Bachelors degree in electrical engineering from the City College of New York (CCNY) and his Masters and Ph.D. degrees from the University of Buffalo in electrical engineering. Before beginning an academic career at Rensselaer Polytechnic Institute, he worked at Bell Aerospace Corporation on the visual simulation of space flight. His specialty is digital signal processing, emphasizing image processing, speech processing, and brain computer interfacing. He conducts sponsored research and teaching in this field, in addition to research in adaptive systems and pattern recognition, and computer integrated manufacturing. He has continuously taught each semester, and is recognized for his teaching excellence, which has included teaching at universities in Europe and Singapore. Lester is a Fellow of both the IEEE and of ASEE and holds several patents. He has served as a Board member of both privately and publically held companies, and is currently is on the Board of Directors of Capintec, Inc.

He has also been actively involved in academic administration, including positions as Department Chair, Founding Director of the Center for Manufacturing Productivity, Director of the Computer Integrated Manufacturing Program, Director of the Center for Industrial Innovation, Associate Dean of Engineering for Research and Strategy, VP of Research Administration and Finance, Dean of Engineering, and, Vice Provost and Dean of Graduate Education. His international work includes positions with NATO and the governments of Singapore, Portugal, Canada, Germany, Ireland, and the United Kingdom, in addition to co-founding the Global EEE and founding the REACH Program at Rensselaer. His honors and awards include: Inaugural Recipient of the National ASEE Research Administration Award; an honorary doctorate from the Technical University of Denmark; Distinguished Alumni Award from the State University of New York at Buffalo; and Senior Advisor to the President of the Institute of International Education (IIE).

JOHN M. GRANDIN

John M. Grandin is Professor Emeritus of German and Director Emeritus of the International Engineering Program at the University of Rhode Island, an interdisciplinary curriculum, through which students complete simultaneous degrees (BA and BS) in German, French, Spanish, or a minor in Chinese, and in an engineering discipline. Grandin has received numerous awards for his work combining languages and engineering, including the Federal Cross of Honor from the Federal Republic of Germany, the Award for Educational Innovation from ABET, and the Michael P. Malone Award for Excellence in International Education from NASULGC, the National Association of State Universities and Land Grant Colleges. He has published widely on such cross-disciplinary initiatives and has been the principle investigator for several funded projects related to the development of the International Engineering Program. Grandin also founded and organized the Annual Colloquium on International Engineering Education, bringing together university faculty and business representatives to develop a more global engineering education nationally(http://uri.edu/iep). Grandin served as associate dean and acting dean of URI's College of Arts and Sciences, and as chair of the Department of Languages. He also has published several articles and a book on Franz Kafka.

E. DAN HIRLEMAN

E. Dan Hirleman is currently professor and dean of the School of Engineering at the University of California, Merced. He earned his degrees from Purdue University, and was a National Science Foundation Graduate Fellow, a Howard Hughes Doctoral Fellow, and an Alexander von Humboldt Foundation Fellow. Hirleman received the 2006 Achievement Award from the International Network for Engineering Education and Research (INEER); ASME Fellow status in 2007; the Hon. George Brown Award for International Scientific Cooperation from the U.S. Civilian Research & Development Foundation (CRDF) in 2008, and the 2009 Charles Russ Richards Memorial Award from Pi Tau Sigma/ASME. His research is in the areas of optical sensors for surface, flow, and biohazard characterization and in global engineering education.

BRENT K. JESIEK

Brent K. Jesiek is assistant professor in Engineering Education and Electrical and Computer Engineering at Purdue University. He holds a B.S. in Electrical Engineering from Michigan Tech and M.S. and Ph.D. degrees in Science and Technology Studies from Virginia Tech. His research examines the social, historical, global, and epistemological dimensions of engineering and computing, with particular emphasis on topics related to engineering education, computer engineering, and educational technology.

JUAN LUCENA

Juan Lucena is Associate Professor at the Liberal Arts and International Studies Division (LAIS) at the Colorado School of Mines (CSM). Juan obtained a Ph.D. in Science and Technology Studies (STS) from Virginia Tech and a M.S. in STS and BS in Mechanical and Aeronautical Engineering from Rensselaer Polytechnic Institute (RPI). His book *Defending the Nation: U.S. Policy making to Create Scientists and Engineers from Sputnik to the 'War Against Terrorism'* (University Press of America, 2005) provides a comprehensive history of the education and development of scientists and engineers in the U.S. in the last five decades. In the 1990s, he researched how images of globalization shape engineering education and practice under a NSF CAREER Award titled *Global Engineers: An Ethnography of Globalization in the Education, Hiring Practices and Designs of Engineers in Europe, Latin America, and the U.S.* He has been co-investigator in projects such as *Building the Global Engineer*, aimed at developing, evaluating, and disseminating curricula on the cultural dimensions of engineering education and practice in different national contexts; *Enhancing Engineering Education through Humanitarian Ethics*, focused on researching and developing curricula at the intersection between 'humanitarianism' and 'engineering ethics'; and *Engineering and Social Justice*, aimed at finding intersections between these two apparently incommensurable fields of practice. Dr. Lucena has served as a member of key advising groups such as NSF/Sigma Xi Steering Committee on U.S. *S&E Globally Engaged S&E Workforce* and NAE's Center for Engineering Ethics and Society. He has directed the Science, Technology, and Globalization Program at Embry-Riddle Aeronautical

University and the McBride Honors Program in Public Affairs for Engineers at CSM. He has been Boeing Senior Fellow in Engineering Education at the National Academy of Engineering, Visiting Professor of Science, Engineering, and Technology Education at the Universidad de las Americas in Puebla (Mexico), and co-editor of *Engineering Studies,* the Journal of the International Network for Engineering Studies.

PHIL MCKNIGHT

Phil McKnight is Professor of German and Chair, School of Modern Languages at the Georgia Institute of Technology. He has published 9 books on East German writers, 18th century studies and satire. He initiated the Georgia Tech/TU Munich/Siemens study/internship program, and manages internships to Japan. He has received numerous major grants from the IIE, Department of Education, DAAD, Fulbright, Japan Foundation and Korean Foundation to develop applied language and intercultural studies opportunities for students at Georgia Tech. Foreign language enrollments have more than doubled since he came to Georgia Tech in 2001. Prior to Georgia Tech he was Chair of the German Department at the University of Kentucky and co-founded the Kentucky-Germany Business Council.

JAMES R. MIHELCIC

James R. Mihelcic is a Professor of Civil and Environmental Engineering and State of Florida 21st Century World Class Scholar at the University of South Florida. He directs the Peace Corps Master's International Program in Civil & Environmental Engineering (`http://cee.eng.usf.edu/peacecorps`). Dr. Mihelcic is a past president of the Association of Environmental Engineering and Science Professors (AEESP) and a member of the Environmental Protection Agency's Science Advisory Board. He is a Board Certified Environmental Engineer Member (BCEEM) and Board Trustee for the American Academy of Environmental Engineers (AAEE). Dr. Mihelcic has traveled extensively in the developing world for service and research. He is lead author for 3 textbooks: *Fundamentals of Environmental Engineering* (John Wiley & Sons, 1999); *Field Guide in Environmental Engineering for Development Workers: Water, Sanitation, Indoor Air* (ASCE Press, 2009); and, *Environmental Engineering: Fundamentals, Sustainability, Design* (John Wiley & Sons, 2010).

D. JOSEPH (JOE) MOOK

Dr. D. Joseph (Joe) Mook is currently a Program Manager in the Office of International Science and Engineering (OISE) at the National Science Foundation, where he is on leave from his permanent position as Professor of Mechanical and Aerospace Engineering (MAE) at the University at Buffalo, State University of New York. He was MAE Department Chair from 2004-2007, and Assistant Dean for International Education for the School of Engineering and Applied Sciences from 1997-2007. He is a recipient of an Alexander von Humboldt Research Fellowship (Germany), and also, a Senior Research Fellowship from the Japan Society for the Promotion of Science. He has been twice elected Chair of the Executive Committee of the Global Engineering Education Exchange (Global E3), and he has also received SUNY system-wide Chancellor's Awards for Excellence in Teaching and for Internationalization. He has been a visiting professor at Darmstadt and Hannover in Germany, at Toulouse, Troyes, and Compiegne in France, and at Chiang Mai in Thailand. His research spans topics in nonlinear optimal estimation, system identification, and controls, has resulted in approximately 100 publications, and he has supervised 13 Ph.D. and approximately 50 M.S. students to completion. He holds B.S., M.S., and Ph.D. degrees in Engineering Mechanics from Virginia Tech.

MICHAEL NUGENT

Michael Nugent is currently Director of the National Security Education Program (NSEP), which supports the Boren Scholars and Fellows Program, a major national effort to increase the quantity, diversity, and quality of the teaching and learning of subjects in the fields of foreign languages, area studies, and other international fields that are critical to the Nation's interests. Dr. Nugent also serves as Director of the Language Flagship, an NSEP program that supports professional level language learning (ILR 3 or higher) for undergraduate students of all majors at U.S colleges and universities. Along with other programs, NSEP's mission is to produce an increased pool of applicants skilled in language and culture for work in the departments and agencies of the United States Government with national security responsibilities.

Before coming to NSEP, Dr. Nugent served in a number of positions supporting international, language and cultural education at the U.S. Department of Education, including Chief of Section, Advanced Training and Research Team, managing Title VI funding of National Resource Centers, Foreign Language and Areas Studies grants, the Language Resource Centers, and the International Studies and Research Grants. He also directed the U.S.-Brazil Higher Education Consortia Program and the North America Mobility in Higher Education Program at the Fund for the Improvement of Postsecondary Education (FIPSE).

Dr. Nugent has served in policy positions as Vice President for Administration and Research at the Council for Higher Education Accreditation (CHEA) in Washington DC, and Deputy to the Chancellor for Systems Relations for Minnesota State Colleges and Universities. Author of "The Transformation of the Student Career: University study in Germany, Sweden, and the Netherlands" (RoutledgeFalmer, 2004), he remains active in the field of international higher education policy. Dr. Nugent has a Ph.D. in higher education from Pennsylvania State University. He has been a student of language and literature at universities in Germany, France, and Spain.

ALAN R. PARKINSON

Alan R. Parkinson was appointed dean of the Ira A. Fulton College of Engineering and Technology in May 2005. Before his appointment as dean, he served as associate dean and also as chair of the Mechanical Engineering department. Dr. Parkinson received his M.S. and Ph.D. degrees from the

University of Illinois and his B.S. and M.B.A. degrees from BYU. His research interests include optimization methods in engineering design, robust design, and engineering education. Dr. Parkinson received the Design Automation Award in 2003 from the American Society of Mechanical Engineers for his work in robust design. He was elected to Fellow status in the American Society of Mechanical Engineers in 2004.

LINDA PHILLIPS

Linda Phillips is a Lecturer and Patel Associate at the University of South Florida in the Civil and Environmental Department. She has a B.S. and M.S. in Civil Engineering from Michigan Technological University specializing in Construction Management. Linda has over twenty years of practical experience working as a project engineer with a large natural gas utility in Michigan and was a project manager and Vice president of Planmark Architects and Engineers, a division of SuperValu Inc., Minneapolis, Minnesota. Linda began her teaching career in 1997 at Virginia Tech and then moved to the University of Minnesota before going to Michigan Tech from 1998 – 2008, teaching classes in Project Management, Professional Practice, and Capstone Design. In 2000, at the request of her students, Linda started the International Senior Design (ISD) taking over 160 students to developing world countries to do their Capstone design projects.

MARGARET PINNELL

Dr. Margaret Pinnell is an associate professor in the Department of Mechanical and Aerospace Engineering at the University of Dayton. She teaches undergraduate and graduate materials related courses including Introduction to Materials, Materials Laboratory and Engineering Design and Appropriate Technology (ETHOS). She is the former faculty director and current associate director for ETHOS (Engineers Through Humanitarian Opportunities of Service Learning) . She has incorporated service-learning projects into her classes and laboratories since she started teaching in 2000. Her research interests include service-learning and pedagogy, K-12 outreach, biomaterials and materials testing and analysis. Prior to joining the School of Engineering, Dr. Pinnell worked at the University of Dayton Research Institute in the Structural Test Laboratory and at the Composites Branch of the Materials Laboratory at Wright Patterson Air Force Base.

ANU RAMASWAMI

Dr. Anu Ramaswami is Professor of Environmental Engineering and Director of the NSF- IGERT Program on Sustainable Urban Infrastructure at the University of Colorado Denver. Ramaswami's research spans environmental modeling, industrial ecology, sustainable infrastructure design, urban systems analysis, and integration of science and technology with policy and planning for sustainable development in communities. Her team is presently working with more than 20 cities in the US and worldwide on understanding sustainability needs, evaluating infrastructure trajectories and prioritizing actions and policies on the ground. She is also leading the development of an inter-disciplinary curriculum on "*Sustainable Infrastructures and Sustainable Communities*" for students drawn from engineering, architecture, planning, public affairs and public health. Since 1996, Ramaswami has advised more than 30 graduate students, co-authored a textbook (on Integrated Environmental Modeling), published more than 50 papers, and managed high-impact applied sustainability research and field projects in communities.

RICK VAZ

Rick Vaz is Dean of Interdisciplinary and Global Studies at Worcester Polytechnic Institute, with responsibility for a worldwide network of undergraduate research programs and an academic unit focusing on local and regional sustainability. His interests include service and experiential learning, sustainable development, and internationalizing engineering education. He held systems and design engineering positions before joining the WPI Electrical and Computer Engineering faculty in 1987.

BERND WIDDIG

Bernd Widdig is Director of the Office of International Programs and Director of the McGillycuddy-Logue Center for Undergraduate Global Studies at Boston College. Before his appointment at BC in 2007, he worked as a professor and administrator at MIT. He joined the faculty of MIT in 1989 and taught courses in German culture and literature as well as seminars on cross-cultural communication. In 1997 he started the MIT-Germany Program, a tailored international program for engineering and science students. In 2001 he was appointed Associate Director of the MIT International Science and Technology Initiative (MISTI). Bernd Widdig has published widely including *Culture and Inflation in Weimar Germany*, Berkeley: University of California Press, 2001 and *In Search of Global Engineering Excellence: Educating the Next Generation of Engineers for the Global Workplace*, Hannover 2006 (co-author). In recognition of his achievements in fostering German-American relations, he received the Cross of the Order of Merit of the Federal Republic of Germany in 2008.

Index